Drought Mitigation and Management

Suresh Kumar
S.P.S. Tanwar
Akath Singh
Central Arid Zone Research Institute (CAZRI),
ICAR,
Jodhpur

SCIENTIFIC
PUBLISHERS

Published by
SCIENTIFIC PUBLISHERS (INDIA)

Jodhpur –

5 A, New Pali Road
P.O. Box 91
Jodhpur - 342 001 INDIA

Delhi –

4806/24, Ansari Road
Daryaganj
New Delhi - 110 002 INDIA

ISBN: 978-93-86102-39-3

Visit the Scientific Publishers (India) website at
http://www.scientificpub.com

Printed in India

Preface

Drought is a natural hazard characterized by lower than expected or lower than normal precipitation, persisting for extended period and is insufficient to meet the demands of human activities and the environment. It has become a perennial malady of rainfed and dry land cropping areas across the globe but it assumes alarming proportion in India. Though considered a normal part of climatic cycle, its spatial extent and severity varies on seasonal and regional time scales. The drought creeps in slowly and imperceptibly resulting in crop failures, depleting water storage in tanks and water bodies, and decline in drinking water availability. All these have far reaching impacts on livelihood of people and survival of livestock thus threatening the social fabric, law and order and destabilizing national economies and international relations. These impacts continue to haunt one and all, directly and indirectly, even long after the drought is over putting immense stress on governments to bring the situation back to normalcy.

The ill effects of drought, to a considerable extent, can be alleviated by adopting proper management strategies. Owing to the frequent occurrence of droughts and their wide spread adverse impacts, both planners and researchers are increasingly engaged in development of strategies to mellow consequent economic, social and environmental impacts. This requires a multidisciplinary approach to deal with the multidimensional problems of drought. Newer approaches for its mitigation for better management of available resources to optimize productivity are being constantly evolved. In fact, research on mitigation and management takes long time to arrive at tangible recommendations that have decidedly favourable impacts. Such updated information on drought and monitoring its vulnerability, adaptability, contingent planning and novel approaches for its mitigation and management have all been painstakingly written and compiled by experienced colleagues in this book and for which we are grateful to them.

The Book contains fundamental as well practical details on drought management and mitigation in agriculture and allied sectors. Each chapter is independently developed from basic conceptualization to advance application and finally case studies and future thrust. These chapters have been grouped into three sections: (1) Basic concepts, assessment, monitoring, forecasting, early warning, vulnerability and adaptation to drought. (2) Mitigation and management strategies. (3) Management of different landuse systems under drought and finally two chapters on socio-economic impacts and livelihood issues with a bearing on policy at both, national and international level. We expect these to be useful to a wide range of stakeholders, i.e., planners, researchers, students and interested public. We hope that this will serve as textbook as well as reference work. It can be a supplementary reading for courses in agronomy, ecology, geography, agrometeorology and remote sensing besides administration and disaster management units in state/central departments of agriculture, forestry, general administration and relief departments in India as well as other countries.

The Editors

Contents

1

Drought : Basic Concepts

K.J. Ananda Kumar[1], Suresh Kumar[2] and S.P.S. Tanwar[3]

INTRODUCTION

Drought is one of the most widespread natural hazards occurring virtually in all the climatic zones of the world. Unlike aridity, which is a permanent feature of climate, drought is a temporary decline in water or moisture content, significantly below the normal or expected amount for a specified period. Such widespread droughts also occur in India, where two thirds of the country's arable area happen to be at one time or the other susceptible to drought; i.e. 68% of the net area sown in country is prone to drought. Out of this 33% is chronically drought prone, receiving rainfall amounting less than 750 mm per annum and 35% drought prone that receive rainfall between 750-1125 mm per annum (Fig. 1, Anonymous 2009). Temporally, drought has occurred at least once in every three years over the last five decades. Since the mid-nineties, prolonged and widespread droughts have occurred in consecutive years, their frequency registering sudden increase in recent times (FAO 2002, World Bank 2003). Due to the growth of population and expansion of agricultural, energy and industrial sectors, the demand for water has increased manifold and even water scarcity has been a regular menace in many parts of the world. This has further been aggravated by other factors, such as climate change and contamination of water supplies.

Various meteorological parameters like temperature, high winds, low relative humidity, timing and characteristics of rains, including distribution of rainy days during crop growing seasons, intensity and duration of rain, and its onset and termination plays a significant role in the occurrence and spread of droughts. Although, a natural hazard, drought differs from other natural hazards in several ways. First, it is difficult to determine the onset and the end of a drought, its

[1]Scientist-D, Ministry of Water Resources, River Development and Ganga Rejuvenation, Govt. of India, New Delhi
[2]Principal Scientist & Ex Head, Central Arid Zone Research Institute, Jodhpur- 342 003
[3]Senior Scientist, Central Arid Zone Research Institute, Jodhpur-342 003

impact is slowly experienced often accumulating over a considerable period and lingering for years after termination. Second, drought impacts are non-structural and spread over larger geographical areas compared to impact resulting from other natural hazards. In contrast to floods, cyclones, Tsunami and earthquakes a drought affects water bodies and seldom results in structural damage. For this reason, the quantification of occurrence of impact and the provision for relief is far more difficult for droughts than for other natural hazards. Third, human activities can directly trigger drought unlike other natural hazards, with exacerbating factors such as excess cropping, excessive irrigation, deforestation, over-exploiting available water, and erosion, adversely impacting the ability of the land to receive and hold water.

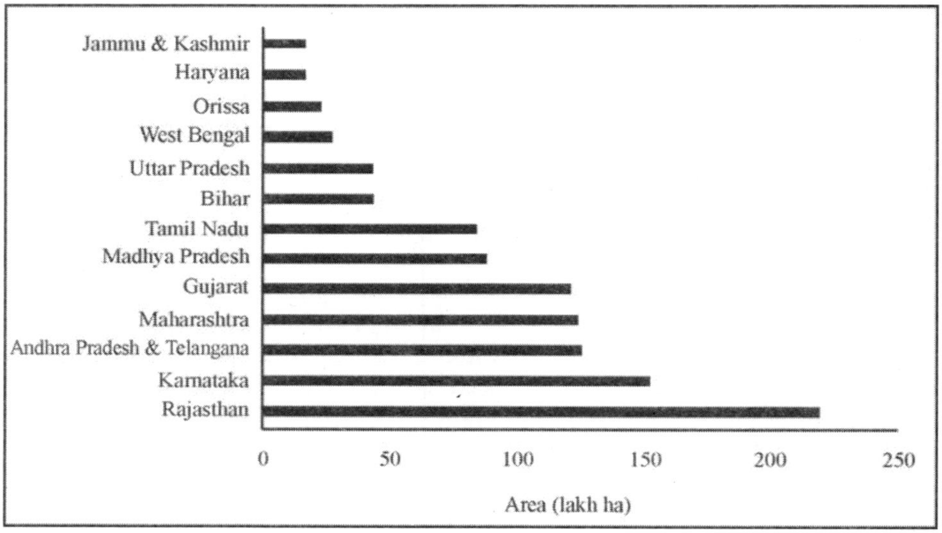

Fig. 1. Extent of drought prone area in different states of India

DEFINITIONS OF DROUGHT

Differences in hydro meteorological variables and socio-economic factors in different regions around the world have become an obstacle in the way of a precise definition of drought. When defining a drought it is important to distinguish between conceptual and operational definitions. Conceptual definitions are those stated in relative terms (e.g., a drought is a long, dry period) whereas, operational definitions on the other hand, attempt to identify the onset, severity, and termination of drought periods. Generally operationally defined droughts can be used to analyze drought frequency, severity, and duration for a given return period. Some of the commonly used definitions are as under:

1. The World Meteorological Organization defines that 'drought means a sustained, extended deficiency in precipitation.'

2. The UN Convention to Combat Drought and Desertification defines it as 'drought means the naturally occurring phenomenon that exists when precipitation has been significantly below the normal recorded levels, causing serious hydrological imbalances that adversely affect land resource production systems.'
3. The Food and Agriculture Organization of the United Nations defines a drought hazard as 'the percentage of years when crops fail from the lack of moisture.'
4. The Encyclopedia of Climate and Weather (Schneider 1996) defines drought as 'an extended period i.e., a season, a year, or several years of deficient rainfall relative to the statistical multi-year mean for a region.'

However, drought definitions vary, depending on the variables used to describe the drought. Hence, drought definitions can be classified into different categories which are discussed below.

TYPES OF DROUGHT

In a given region droughts can be classified into four categories.

1. **Meteorological Drought**: When there is prolonged absence or marked deficiency or poor distribution of precipitation below a given normal value, it is called meteorological drought. As per Indian Meteorological Department meteorological drought is categorized as

   Mild Drought : a. Negative 25% deviation from normal rainfall

   Normal Drought : b. Negative 50% deviation from normal rainfall

   Severe Drought : c. Negative 75% deviation from normal rainfall

2. **Hydrological Drought**: When meteorological drought persists for a prolonged period, there is a marked depletion of surface water with dried up lakes, rivers and reservoirs. This situation where reduction in water resources falls below a specified level for a given period of time, it becomes a hydrological drought.
3. **Agricultural Drought**: It refers to an extended dry period in which the lack of rainfall results in insufficient soil moisture in the crop root zone, causing adverse effects on arable crops. Several drought indices, based on a combination of precipitation, temperature and soil moisture, have been derived to study agricultural droughts.
4. **Socio-economic Drought**: This occurs when the social fabric of a society is affected in terms of increase in unemployment, distress migration, etc. which might have happened due to the disturbed economy of a region caused by extreme meteorological and hydrological droughts.

IMPACTS OF DROUGHT

Drought impacts both surface and groundwater resources and can lead to reduced water supply, deteriorate water quality, crop failure, reduced range productivity, diminished power generation, disturbed riparian habitats, and suspended recreation activities, as well as impact a host of economic and social activities (Riebsame et al 1991). The impact of drought varies with the time scale of droughts. The longer the period of drought and the larger the number of consecutive droughts, the greater will be its impact on agriculture, ecology and economy of that area. These phenomena are summarized by National Drought Monitoring Centre (NDMC), University of Nebraska that is pertinent to Indian condition as mentioned in Tables 1, 2, 3.

Table 1: Environmental impact of drought

	Direct Impact	Indirect Impact
Hydrological	• Lower water level in reservoirs, lakes and ponds • Reduced stream flow • Loss of wetlands • Increased ground-water depletion and land subsidence	• Increased time and cost for water collection and transfer • Lower water quality eg. Salinization and temperature increase • Water borne diseases • Wind and water erosion on soil
Biological	• Loss of trees and vegetation • Loss of animal species diversity	• Fragmentation and destruction of wildlife habitats • Migration, concentration and increased predation • Loss of biodiversity

Table 2: Economic impacts of drought

	Direct Impact	Indirect Impact
Cost and losses to agricultural producer	• Annual and perennial crops losses • Damage to crop quality • Reduction in productivity of cropland , e.g., wind erosion • Insect infestations • Plant diseases • Wildlife damage to crops	• Income loss to farmers because of reduced crop yield • Increased irrigation costs • Cost of new or supplemental water resources development, e.g., wells, dams and pipelines • Long-term loss to industries directly dependent on agricultural production, e.g., food processors • Increased commodity prices
Cost and losses to	• Reduced productivity of range land, animal carrying capacity	• Forced reduction of foundation stock (seed)

livestock producer	• Increased travel time for grazing • Decease stock weights and reduced milk production • Increased livestock diseases • Closure/limitation of public lands to grazing • Range fires	• High cost/unavailability of feed or water for livestock • Reduction in livestock market price • Increased feed transportation costs • Disruption of reproduction cycle (delayed breeding, more miscarriages) • Increased predation and pouching
Cost and losses to industry and urban activities	• Higher cost of water and sanitation • Decreased in public water supplies • Impacts on transportation • Higher cost/lower availability of hydro-electric power	• Higher cost or unavailability of water for horticulture, agri food processing and value-added manufacturing • Impaired productivity of forest land and reduced timber production • Increased pollution e.g., dust • Increased diseases • Reduction in tourism revenue, e.g., wildlife • Strain on financial institution, e.g., greater credit risks

Table 3: Social impacts of drought

	Direct Impact	Indirect Impact
Reduced quality of life	• Increased work load for women in collection of fuel-wood and water • Reduced levels and variety of food sources • Increased government expenditure on relief	• Increased poverty • Migration (rural to urban areas, cross-border) • Reduction or modification of recreational activities • Disruption of cultural practices and belief/value system • Loss of cultural sites and aesthetic values
Increased conflicts		• Water user conflicts • Political conflicts • Management conflicts • Other social conflicts, eg., scientific and media based
Health	• Physical and emotional stress, eg. anxiety, depression and	• Health related low-flow problems, e.g., cross- connection contamination, diminished sewage flows, increased

	loss of security	pollutant concentration and reduced fire fighting capability • Reduction in nutrition • Loss of human life • Public safety from forest and range fires • Increased respiratory ailments • Increased health hazards

Drought Management

As the drought re-occurs, management strategies for it either at regional or individual level, should be at the sustainable basis rather than looking for a crisis management approach (Fig. 2). There should be comprehensive, long-term drought preparedness policies and plans of actions that may significantly marginalize the risks and vulnerabilities to drought. Some of the important components of drought management strategy are:

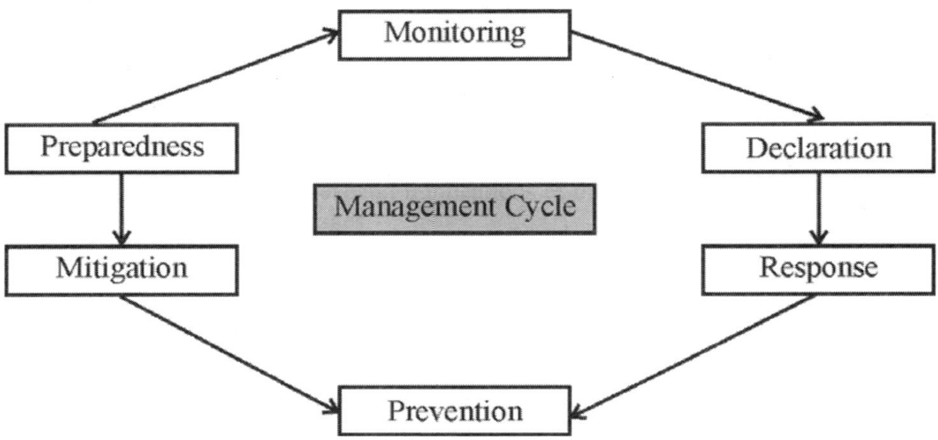

Fig. 2. Drought management cycle

Drought Assessment

- Drought prediction
- Detection
- Developing early warning system
- Monitoring drought scenerio- spread, withdrawl.
- Vulnerability assessment.
- Impact assessment
- Mid term correction

Drought Management

- Adoption and response
- Management of available natural resources
- Cropping strategies
- Contingency crop planning
- Drought policy including immediate, medium and long term perspectives
- Relief measures for mitigation of drought impacts

Indicators of Drought

Above management strategies require a fairly good understanding of drought indicators. But in view of the complexity of drought variability according to climatic and geographic conditions, it seems appropriate to work on different parameters to be included in local or national indicators that could be calibrated and compared, when sufficient data is available. The presence or absence of these parameters in local indicators will depend on their local relevance. Some of the common indicators are:

- Persistently deficient rains accompanied with long dry spells
- Low water storage in reservoirs
- Drying up of wells, tanks and other water bodies
- Wilting stages of crops

Collaborating Agencies for Drought Management in India

- Ministry of Earth Sciences
    - India Meteorological Department
    - National Centre for Medium Range Weather Forecasting
    - Indian Institute of Tropical Meteorology
- Indian Council of Agricultural Research
- Department of Agriculture & Cooperation
- State Departments of Agriculture
- State Agricultural Universities and other Universities
- Ministry of Information Technology
- Ministry of Science & Technology
- Ministry of Information & Broadcasting (AIR & TV)
- Print Media
- Department of Space
- Ministry of Rural Development

- Ministry of Water Resources
- NGOs and others

EPILOGUE

Modernization has made it possible for people to effectively mitigate much of the impact of drought through irrigation and crop rotation. Nation's economic growth is sustainable only if there is a long term plan to deal with monsoon failure and water scarcity. The rural community should be part of planning and development of drought proof strategies so that the plans can be implemented on ground with assured results to deal with monsoon failure and water scarcity.

REFERENCES

Anonymous 2009 *Manual for Drought Management*. Department of Agriculture & Cooperation. Ministry of Agriculture, New Delhi.

FAO 2002 Report of FAO-CRIDA Expert Group Consultation on Farming System and Best Practices for Drought-prone Areas of Asia and the Pacific Region. Food and Agricultural Organisation of United Nations. Published by Central Research Institute for Dryland Agriculture, Hyderabad, India.

Riebsame WE, Changnon SA, Karl TR 1991 Drought and Natural Resource Management in the United States: Impacts and Implications of the 1987–1989 Drought. Westview Press, Boulder, CO, 174 pages

Schneider SH 1996 Encyclopaedia of Climate and Weather. Oxford University Press, New York.

World Bank 2003 Report on Financing Rapid Onset Natural Disaster Losses in India: A Risk management Approach. Report No. 26844-IN, Washington, DC.

2

Drought – Indicators, Indices and Monitoring Checklist

R.K. Goyal[1*], K.J. Ananda Kumar[2] and S.P.S. Tanwar[3]

INTRODUCTION

Drought is one among the natural disasters that weakens the sustainable development of people and their resources. Generally it is a period of drier than normal conditions that results in water related problems. Drought affects soil moisture, flow in streams, drop in level of reservoirs and groundwater table etc. (Fig. 1). Drought onset and end are often difficult to determine, as is its severity. The impacts of drought are largely non-structural and spread over a larger geographical area than are damages from other natural hazards. The non-structural characteristic of drought impacts has certainly hindered the development of accurate, reliable, and timely estimates of severity and, ultimately the formulation of drought preparedness plans by most governments. Though drought is a normal climatic phenomenon that occurs in all climatic zones. The intensity and frequency of drought usually increases towards drier climatic zones. Depending on the severity, drought causes different degree of sufferings to the human society, especially those depending more on agriculture and allied activities. Drought can also be classified depending upon it's time of occurrence, duration and likely people affected by drought. The impacts of drought, like those of other hazards, can be reduced through mitigation and preparedness.

DROUGHT INDICATORS

Indicator is a measure of meteorological, hydrological, agricultural, or socio-economic variable that provides an indication of potential drought related stress

1. Principal Scientist (SWCE), Central Arid Zone Research Institute, Jodhpur 342 003
2. Scientist-D, Ministry of Water Resources, River Development and Ganga Rejuvenation, Govt. of India, New Delhi
3. Principal Scientist, Central Arid Zone Research Institute, Jodhpur 342 003
*E-mail: rkgoyal24@rediffmail.com

or deficiency. It enables people to describe drought in a consistent way across different times and places. These indicators combine variables such as precipitation and temperature. Some indicators represent drought within specific intervals of time.

1. Rainfall

The Indian Meteorological Department (IMD) and State Governments collect data on rainfall every day during the rainy season. The IMD maintains its network of weather stations throughout the country. Within the state, data are collected at the tehsil /taluka/block level(s). Generally, the Revenue Department is entrusted with the responsibility of collecting rainfall data at the district and tehsil levels. The actual rainfall is compared with the Long-Term Average (LTA), which has been standardized on a daily, weekly and monthly basis. Such a comparison provides information on the deficit or excess of rainfall in a particular sub-division for a certain period. According to the IMD, drought sets in when the deficiency of rainfall at a meteorological sub-division level is 25 per cent or more of the LTA of that sub-division for a given period. The drought is considered "moderate'; if the deficiency is between 26 and 50 per cent, and "severe" if it is more than 50 per cent.

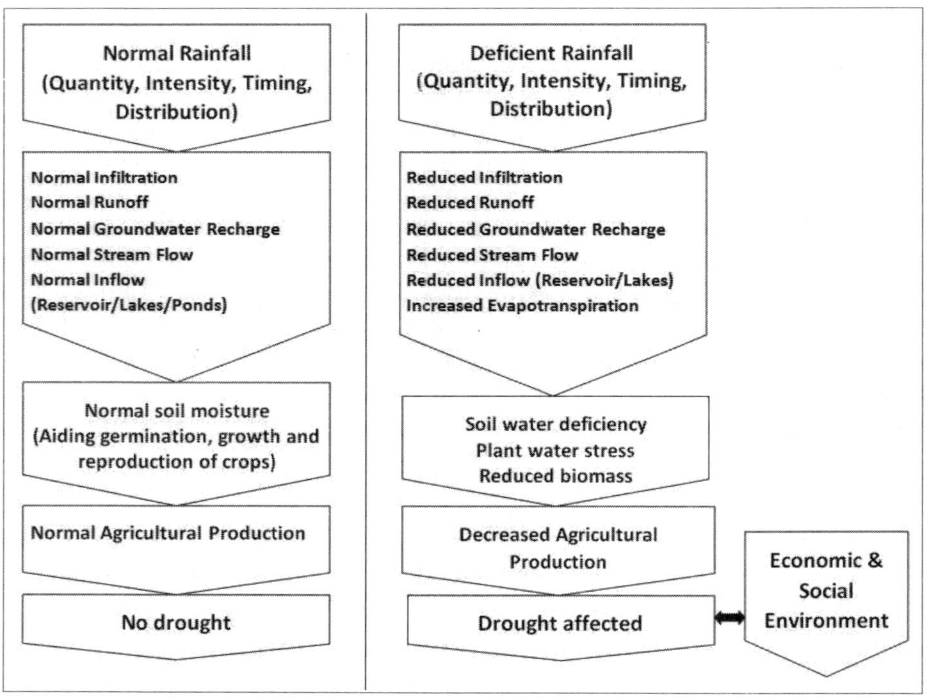

Fig. 1. Drought cycle (NIDM 2009)

2. Storage Water Levels in Reservoirs

State Governments collect data on the levels of stored water in important reservoirs through its Irrigation Department. Reservoir storage level is useful indicator of water shortages. As data on reservoir storage are available on a regular basis, these could provide accurate information on water shortages. The Central Water Commission maintains data on water levels in 81 important reservoirs of the country, where the water storage is compared with the 'full Reservoir Level' (FRL). State Governments need to plan the use of reservoir storage as per their reservoir operation rules, which lay down the priority for the use of available water: drinking water, urban and industrial use and irrigation for agriculture. Reservoir water levels may fall below the expected level even when the reservoir operation rules are not followed and the usage of water is not adequately regulated.

3. Surface Water and Groundwater Level

Natural discharge from shallow aquifers provides base flow to streams and sustains the water in lakes and ponds, particularly during periods of dry weather. Similarly, groundwater levels are also affected due to poor recharge, either due to lack of adequate rainfall or poor water conservation practices. As a result, water availability in deep bore-wells and open wells diminishes substantially. Declining groundwater levels are important indicators of drought conditions, though these are often attributed to over extraction of water. An annual decline in the water table of up to 2 meters is considered normal and can tolerate even a deficient rainfall the following year. A decline of up to 4 meters is a cause for concern and above 4 meters is a stress situation.

The Central Ground Water Board (CGWB), with over 15,000 hydrograph stations across the country is responsible for monitoring India's groundwater. Monitoring, usually done four times a year, is essentially a recording of the response of the groundwater system to natural and artificial conditions or recharge and discharge. Most of the State Governments have similar groundwater boards or agencies for surveying groundwater levels and their periodical reports provide information upon declining groundwater levels.

4. Sowing and Crop Conditions

An important indicator of drought is the total area sown. The State Government agriculture department provides information on sowing on a weekly basis. A delayed sowing shows rainfall deficiency and indicates the onset of drought. Reports on crop conditions also provide an indication of the severity of the drought situation. If the crops are wilting, it indicates soil moisture stress. A crop contingency plan and other mitigation measures are implemented based on reports prepared for all the crops sown during the monsoon.

DROUGHT INDICES

Drought index is a method of deriving "value added" information related to drought by comparing current conditions to historical information based upon statistical calculations. Commonly, a drought index is a prime variable for assessing the effect of a drought and defining different drought parameters, which include intensity, duration, severity and spatial extent. A number of different indices have been developed to quantify a drought, each with its own strengths and weaknesses (Mishra and Singh 2012). The IMD carry out drought monitoring mainly based on deviation of rainfall from normal, aridity anomaly index and standardized precipitation index. Practically all drought indices use precipitation either singly or in combination with other meteorological elements, depending upon the type of requirements. Some commonly used drought indices are discussed here.

1. Aridity Anomaly Index

The IMD has developed an Aridity Anomaly Index based on rainfall, potential evapotranspiration and actual evapotranspiration taking into account soil moisture conditions and using the water budgeting method. Aridity anomalies are worked out based on this index and these anomalies are classified into various categories of arid conditions i.e. *Mild Arid* (aridity anomaly 1-25%), *Moderate Arid* (aridity anomaly 26-50%), and *Severe Arid* (aridity anomalies more than 50%). These anomalies are used for near real-time monitoring and assessment of agricultural droughts across the country at weekly/fortnightly intervals. This indirectly helps to assess the moisture stress experienced by growing plants.

2. Standardized Precipitation Index

The Standardized Precipitation Index (SPI) is a relatively new drought index based on long term precipitation record for a desired period (McKee et al 1993, 1995). The SPI assigns a single numeric value to the precipitation, which can be compared across regions and time scales with markedly different climates. This spatial and temporal flexibility allows the SPI to be useful in both short-term agricultural and long-term hydrological applications. Since the SPI expresses the possibility of simultaneously experiencing wet conditions on one or more time scales and dry conditions at other time scales, it is often a difficult concept to convey in simple terms to decision-makers.

3. Palmer Drought Severity Index

Palmer Drought Severity Index (PDSI) uses precipitation and temperature for estimating moisture supply and demand within a two-layer soil model (Palmer 1965). PDSI is perhaps the most widely used regional drought index for monitoring droughts globally. PDSI varies roughly between -6.0 and +6.0. More wet conditions are indicated by positive values of PDSI, while more dry conditions show negative values. Thresholds for classification of wetness are arbitrary. PDSI values between -2 and +2 would indicate normal conditions, although the sub-range of -1 to -2 could also be treated as mild drought. PDSI values in the range of -2 to -3 are indicative of moderate drought, -3 to -4 indicate severe drought and values below -4 would be associated with extreme drought. PDSI values are normally calculated on a monthly basis. Further interpretation of monthly PDSI allows drought duration to be taken into account as well. A drought sequence is interpreted as a sequence of 3 or more consecutive months with a PDSI value $\leq$ -2.0. A series of 6 or more months is a major drought event. The end of a drought sequence is taken as the last month where the PDSI is $\leq$ -2.0.

4. Crop Moisture Index

The Crop Moisture Index (CMI) complements the PDSI. It measures the degree to which crop moisture requirements are met, is more responsive to short-term changes in moisture conditions and is not intended to assess long-term droughts (Palmer 1968). CMI is normally calculated with a weekly time step and is based on the mean temperature, total precipitation for each week and the CMI value from the previous week. For each growing season, CMI typically begins and ends near zero. It is, in principle, possible to use a combination of PDSI and CMI for drought monitoring, where PDSI would serve as a long-term drought-monitoring tool and the CMI may indicate the progression of seasonal water shortages during a crop growing stage.

5. Surface Water Supply Index

The Surface Water Supply Index (SWSI) integrates reservoir storage, stream flow, and two precipitation types (snow and rain) at high elevations into a single index number (Shafer and Dezman 1982). SWSI is relatively easy to calculate and it gives a representative measure of water availability across a river basin or selected region/province. It is, however, unlikely that it could be successfully used for large regions with significant spatial hydrological variability: the weights may differ substantially from one part of the region to another.

6. Normalized Difference Vegetation Index

Normalized Difference Vegetation Index (NDVI) is based on the concept that vegetation vigour is an indication of water availability or lack thereof. It shows us the effects of climate on vegetation in terms of its absorptive capacity in visible light but little in the near-infrared. The difference of visible and near-infrared reflectance represents photosynthetically active vegetation; this information is used to construct a vegetation index. The lowering of the vegetation index indicates moisture stress in vegetation, resulting from prolonged rainfall deficiency. High NDVI values might reflect ideal climate growing conditions so that vegetation greenness is higher than encountered in other years.

7. Normalized Difference Wetness Index

Normalized Difference Wetness Index (NDWI) is expected to give the vegetation or crop turgidity and health. It is based on the use of Shortwave Infrared (SWIR) band, which is sensitive to moisture available in soil as well as in crop canopy. In the beginning of the cropping season, soil background is dominant which makes SWIR sensitive to soil moisture in the top 1-2 cm. As the crop progresses, SWIR becomes sensitive to leaf moisture content. NDWI using SWIR can complement NDVI for drought assessment particularly in the beginning of the cropping season. Higher values of NDWI signify more surface wetness.

8. Effective Drought Index

Unlike many other drought indices, the Effective Drought Index (EDI) in its original form is calculated with a daily time step. EDI is a function of precipitation needed for a return to normal conditions (PRN). PRN is precipitation, which is necessary for the recovery from the accumulated deficit since the beginning of drought. PRN, in turn, effectively stems from daily effective precipitation (EP) and its deviation from the mean for each day.

9. Moisture Adequacy Index

The Moisture Adequacy Index (MAl) is obtained from weekly water balance. Drought impact is related to moisture availability at certain crop growth stages. Hence, categories of MAl (severity) at different growth stages are integrated into a single index value to identify drought impact on a particular crop. The Central Arid Zone Research Institute (CAZRI), Jodhpur monitors agricultural drought in the Indian arid regions by using MAl.

DROUGHT MONITORING CHECKLIST (NIDM 2009)

Meteorological Data

Agencies: India Meteorological Department, National Centre for Medium Range Weather Forecasting, State Governments

Indices to be monitored: Daily, weekly, and monthly rainfall, snow fall/fog

Hydrological Data

Agencies: Central Water Commission, Central Ground Water Board, State Governments (irrigation departments, groundwater agencies, water resources departments/ projects)

Indices to be monitored: Water storage in reservoirs / ponds / lakes, river flow, groundwater level, yield and draft from aquifers, water loss through evaporation, leakage, seepage.

Agricultural Data

Agencies: National Crop Forecasting Centre, Directorate of Economics & Statistics, Indian Council of Agricultural Research, Agricultural Census Data, State Government agriculture departments, agricultural universities

Indices to be monitored: Soil moisture, area under sowing and type of crop, crop water requirement, status of growth, crop yield, alternative cropping possibilities, Land holdings

Data from Space

Agencies: National Remote Sensing Centre

Indices to be monitored: Vegetation monitoring, rainfall, surface wetness and temperature monitoring.

Socio-economic Data

Agencies: Planning Commission, Department of Food & Public Distribution, Department of Consumer Affairs, Department of Rural Development. Ministry of Women and Child Development Department, of Animal Husbandry, Dairying & Fisheries, Revenue Department of state etc.

Indices to be monitored: Availability and prices of foodgrains, availability of fodder, migration of population, distress sale of assets, vulnerable areas / population

CONCLUSION

There are many challenges before us if we are to improve our management of droughts. First, drought must be accepted as a natural hazard within the natural hazard community of scientists and policy makers. Because of its slow-onset characteristics and lack of structural impacts, it is often given less attention. This lack of recognition of the importance of drought by the natural hazards community has been an impediment to obtaining adequate research support and, in many instances, an obstacle to building awareness among policy makers at the local, national, regional, and international level. This lack of awareness in turn has resulted in an under appreciation of drought and its far-reaching impacts. It has also perpetuated the process of dealing with drought in a crisis management mode when the knowledge and technology necessary to improve preparedness and mitigation impacts is readily available. A second challenge is to build awareness of drought as a normal part of climate. It is often considered to be a rare and random event–thus the lack of emphasis on preparedness and mitigation.

Improved understanding of the different types of drought and the need for multiple definitions and climatic/water supply indicators that are appropriate to various sectors, applications, and regions is a critical part of this awareness-building process.

REFERENCES

McKee TB, Doesken NJ, Kleist J 1993 The Relationship of Drought Frequency and Duration to Time Scales, Paper Presented at 8th Conference on Applied Climatology American Meteorological Society, Anaheim, CA

McKee TB, Doesken NJ, Kleist J 1995 Drought Monitoring with Multiple Time Scales, Paper Presented at 9th Conference on Applied Climatology, American Meteorological Society, Dallas, Texas.

Mishra AK and Singh VP 2012 A review of drought concepts, *Journal of Hydrology* **391** 202–216

NIDM 2009 Manual for Drought Management. Developed by National Institute of Disaster Management. Department of Agriculture and Cooperation, Ministry of Agriculture, Government of India, New Delhi 190 pages

Palmer WC 1965 Meteorologic Drought. US Department of Commerce, Weather Bureau, Research Paper No. **45**, 58 pages

Palmer WC 1968 Keeping track of crop moisture conditions, nationwide: the new crop moisture index, Weather wise **21** 156–161

Shafer BA, Dezman LE 1982 Development of a Surface Water Supply Index (SWSI) to Assess the Severity of Drought Conditions in Snowpack Runoff Areas In: Preprints, Western Snow Conf. Reno NV Colorado State University, 164-175.

3

Monitoring and Early Warning System in Drought Management

A.T. Jeyaseelan

INTRODUCTION

Drought is a regional phenomenon and its characteristics differ from one climate regime to another. Impacts are also regional in nature, reflecting exposure to the hazard and the vulnerability of society to extended periods of precipitation deficits. Impacts are a measure of vulnerability. Risk is a product of exposure to the hazard and societal vulnerability. Drought by itself is not a disaster. Whether it becomes a disaster depends on its impact on local people, economies and the environment and their ability to cope with and recover from it. Therefore, the key to understanding drought is to grasp its natural and social dimensions. The goal of drought risk management is to increase society's coping capacity, leading to greater resilience and a reduced need for government's interventions in the form of disaster assistance. A critical component of drought risk management is a comprehensive drought monitoring system that can provide early warning of drought's onset and end, determine its severity and deliver that information to a broad clientele in many climate- and water-sensitive sectors in a timely manner. With this information, the impacts of drought can be reduced or avoided in many cases. In this chapter, the Monitoring and Early warning system for drought management in India and in other parts of the world is discussed.

DROUGHT INDICATORS

Drought indicators can be identified from the types and impact of drought. There are four types of drought namely meteorological, hydrological, agricultural and socio-economic. In the Meteorological drought, precipitation deficiency in terms of amount, intensity and timing and the weather factors such as high temperature, high winds, low relative humidity, greater sunshine, less cloud cover that

Scientist Engineer SG - RCNRSC, Balangar, Hyderabad - 500 072
E-mail: jeyaseelan.at@nrsc.gov.in

increases evaporation and transpiration are the major indicators of drought. In the Hydrological drought, the reduced infiltration, runoff, deep percolation and ground water recharge leads to reduced stream flow, inflow to reservoirs, lakes and ponds, reduced wetlands and wildlife habitat are the major indicators of drought. In the Agricultural drought, reduced soil moisture leads to plant water stress, reduced biomass and yield are the major indicators of drought. In the Socio-economic drought, the scarcity of essential commodity such as food, water, fodder and electricity and the low purchasing power due to reduced income are the major indicators of drought. The sequence of impacts associated with meteorological, agricultural, and hydrological drought also indicate the drought (Figure 1). For example when the drought begins, the agricultural sector is usually the first to be affected because of its heavy dependence on stored soil water. Soil water can be rapidly depleted during extended dry periods. If precipitation deficiencies continue, then people dependent on other sources of water will begin to feel the effects of the shortage.

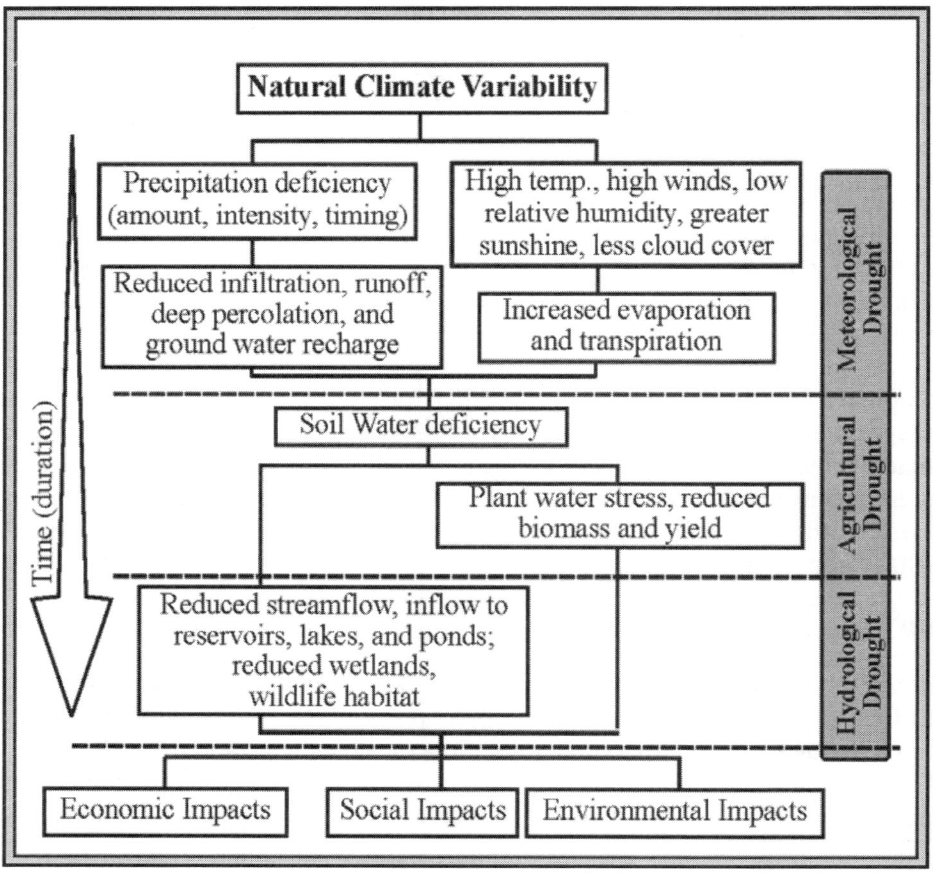

Fig. 1. Concepts of drought

DROUGHT MONITORING AND EARLY WARNING SYSTEM IN INDIA

No single definition of drought works in all circumstances, so water planners rely on indicators or data to recognize droughts. Few indicators appear broadly, sequentially reflecting the extent and depth of household stresses. Agro-climatic indicators are leading indicators that reflect the initial signs of risk minimization strategies adopted by households. Market socioeconomic indicators are concurrent and denote risk absorption strategies. Late anthropometric indicators reflect survival symptoms. Therefore, the Drought Monitoring activity relies on such indicators. Some of the indices used for drought monitoring are: Palmer Drought Index, Crop Moisture Index, CPC Soil Moisture Model (percentiles), Daily Stream flow (percentiles), Percent of Normal Precipitation, Topsoil Moisture (percent short and very short), and the remotely sensed Satellite Vegetation Health Index.

The objective of designing an early warning system is to keep track of leading indicators to get ample lead-time to intervene at the drought onset phase itself, compared to early warning systems designed to capture concurrent or late indicators. However, most interventions based on late indicators force governments to adopt a crisis management approach to dealing with drought-induced food insecurity stresses. There are many deficiencies in this approach; and in the long-run it does not reduce vulnerability to drought.

Early Warning Systems in India

Studies of Indian drought management approaches in the last hundred years reveal that India relied too heavily on crisis management approaches before and during the pre-independence era (before 1947). However, after the mid-sixties, based on the experience of tackling the 1966 drought-induced food crisis, serious efforts were made to replace adhoc crisis management on relief interventions with an anticipatory drought management approach. A drought management approach differs from a relief approach with regard to objectives, reliance on early warning indicators and timing of public intervention. Thus the drought management approach aims at ensuring food production, relying on leading agro-climatic indicators, like rainfall, water level in reservoirs and progress of cropping pattern, to detect early signs of a developing drought situation.

A national-level crop weather watch group (drought forecasting) was established in 1979 to monitor the impact of the monsoon on crop conditions and to suggest corrective measures to minimize crop losses. The National Early Warning System established at India integrates both drought forecasting and monitoring system. While a drought relief approach with a monitoring system enabled the government to intervene only in the months of November- December after the kharif harvest (summer crops) was over, the drought management approach with a forecasting system enabled the government to intervene in July-August (within the monsoon season). This early warning system offered a lead-time of five months before the appearance of distress indicators in December-January. This was evident in the management of a severe multi-year drought during 1985-87.

There are two components of the National Early Warning System: drought forecasting and drought monitoring. The drought forecasting function is carried out by the Inter-Ministerial National Crop Weather Watch Group (CWWG), which meets during the monsoon period from June to September. It monitors the impact of the monsoon on agricultural operations and also suggests corrective measures to minimize any possible adverse impact of aberrant monsoon conditions on crop production as per the standing contingency crop plan. In case the CWWG anticipates widespread adverse seasonal conditions, it sends out a report. This triggers the operationalization of an emergency contingency action plan for drought management, which envisages institutional arrangements and operating procedures for the drought monitoring system.

Breaks in the monsoon rains can be of different durations. Breaks of shorter duration, like 5 to 10 days, may not be of serious concern. But prolonged breaks of more than 2 weeks can create plant water stress, leading to low productivity of crops. These breaks cannot be predicted in advance. The agricultural droughts resulting from prolonged breaks in the monsoon rains can be of different magnitudes and severity and affect different crops in varying degrees. Meteorological information, in terms of the frequency and probability of these breaks, can be used to select a combination of crops of different durations in such a way that there is a time lag in the occurrence of their growth for appropriate intercropping systems.

Long-range seasonal forecast of the Indian Meteorological Department

The Indian economy is basically an agro-based economy. The nation's agricultural planning is primarily dependent on the reasonable accurate prediction of the total amount of rainfall from the beginning of June to the end of September. This kind of prediction comes under the category of long-range forecast (LRFs). The first operational LRF for seasonal monsoon rainfall was issued on June 4, 1886. This was based on antecedent Himalayan snow cover (Blanford, 1884). Over the last hundred years, there have been many refinements in the LRF techniques, including the multiple regression technique introduced by Sir Gilbert Walker in 1910. Presently, IMD issues LRFs based on statistical techniques. In the last 25 years (1988-2013), all the monsoon rainfall forecasts have proved to be fairly accurate. Due to large variation in 2002, a new two model approach has initiated since 2003. The first model uses 8 parameters observed by March and provide probability of the monsoon rainfall by April 16 instead of May 25. The second model uses 10 parameters observed by June and provides forecast update by mid-July (www.imd.gov.in).

Monitoring and Early Warning Services rendered by the Agrimet division of IMD

A scientific study of the influence of weather on crops is of vital importance. Any abnormalities in the weather during the season, such as delay in the outbreak of rains, untimely or excessive rains, droughts, or spells of too-high or too-low temperatures, would very seriously affect the growth and final yield of the crops. Analysis of the existing data indicates that at least 50% of the variability of crop yields is related to weather.

Realizing the importance of meteorology in better understanding the crop-weather relationship and thereby increasing the food production in the country, the Agricultural Meteorology Division (Division of Agricultural Meteorology) was started in August 1932 in Pune. The Division of Agricultural Meteorology has a wide network of agrometeorological observatories and taken up more specific research problems like water requirement of crops, pests and diseases, rainfall probabilities in the dry farming tracts, crop-weather relationship, and application of remote sensing techniques in agricultural meteorology etc.

The Drought Research Unit was set up at IMD in Pune under the instruction of the Planning Commission of the government of India in June 1967. IMD identifies meteorological drought for subdivisions every year based on rainfall analysis. During the last 138 years (1875-2013), IMD has identified meteorological droughts (moderate or severe) over meteorological subdivisions of the country using IMD criteria and also drought years for the country as a whole.

IMD monitors agricultural drought once every two weeks on a real-time basis during the main crop seasons (Kharif and Rabi) of India. For this, an aridity index (AI) developed on the lines of Thornthwaite's concept is used to monitor the incidence, spread, intensification, and recession of drought. AI is given as

$$AI = \frac{PE - AE}{PE} \times 100$$

where PE is potential evapotranspiration calculated with the help of Penman's formula, which takes into account mean temperature, incoming solar radiation, relative humidity, and wind speed. AE is actual evapotranspiration calculated according to Thornthwaite's water balance technique, taking into account PE, actual rainfall, and field capacity of the soil.

The Aridity Anomaly Index (AAI) is calculated by using the normal Aridity Index for 210 well-distributed stations over the country. The Aridity Anomaly Areas are demarcated in Table 1. With the help of aridity anomalies, crop stress conditions in various parts of the country are being monitored during the monsoon season. These anomalies are being used for crop planning and in the early warning system during drought/desertification.

Table 1: Demarcation of areas based on aridity anamoly index

Aridity anamoly Index (AAI)	Aridity Anomaly Areas
0 or negative	Non-arid
1 – 25	Mild arid
26-50	Moderate arid
> 50	Severe arid

The Drought Research Unit also provides **Crop Yield Forecasts** (CYFs). This unit has developed pre-harvest crop yield forecasting models and issues of monthly state wise crop yield and countrywide total production forecasts for the major crops of *kharif* (rice) and Rabi (wheat), based on the agrometeorological models and weather parameters like rainfall, temperature, relative humidity, cloud amounts, and yield. Based on a long series of past crop yield data and meteorological data of the corresponding period, pre harvest crop yield forecasting models have been developed using the multiple regression technique. Pre-harvest crop yield forecasts are issued for 15 states comprising 26 meteorological subdivisions for kharif (rice) and 12 states comprising 16 meteorological subdivisions for rabi (wheat) and also for the total rice/ wheat production of the country. The forecasts are supplied to India's Directorate of Economics and Statistics, Ministry of Agriculture. The first interim forecast for kharif rice is issued in August and the final forecast is given in November/ December. For wheat, the first interim forecast is issued in January and the final in March/ April/May.

Agrometeorological Advisory Services Unit (AASU): The Division of Agricultural Meteorology, in coordination with the respective state agricultural departments, is issuing weekly/bi-weekly Agromet Advisory Bulletins from 17 AAS units located at State Meteorological Centres (MCs)/Regional Meteorological Centres (RMCs). The scheme was first launched in 1977 from the state of Tamil Nadu. The Advisory Bulletins contain specific agricultural advisories tailored to the needs of the farming community. The first and foremost aim of the service is to render timely advice on the actual and expected weather and its likely impact on the various day-to-day farming operations. Short-range forecasts valid for 12 to 24 hours and then extended to the following 2 to 3 days are used extensively to provide this advice. Secondly, agrometeorological forecasts extending over a week or 10 days (medium range) are very important from the users' point of view as well as for planning for various agricultural operations and strategies. The advisories are prepared by taking into account the stage of the crops, agricultural operations in progress, prevalence of pests and diseases, and the immediate impact of weather on crops. They are prepared in consultation with the experts of the State Department of Agriculture and broadcast over All India Radio (AIR) stations, TV network in the state. These bulletins are also sent to various agricultural authorities in the state.

In addition to this bulletin, the Farmers' Weather Bulletin (FWB) is also regularly issued from all MCs/RMCs. Weather services for the farmers in India were started by the India Meteorological Department in 1945. FWBs indicate the onset of rains, probable rainfall – intensity and duration, weak or a break in monsoon conditions, occurrence of frost, hail, squalls, and other conditions. Since 1990, FWBs have been issued throughout the year twice a day. These bulletins and the daily weather reports prepared by the centers are also published in the newspapers. At present, 70 radio stations broadcast FWBs in India. The validity of these forecasts is for 48 hours, with an outlook for the subsequent two days. Contents of the Bulletin are (1) a summary of past weather, (2) a district forecast of weather during the next 48 hours and special weather warnings for cyclonic storms, etc., and (3) an outlook for the subsequent two days (www.imdagrimet.gov.in).

The National Centre for Medium Range Weather Forecasting (NCMRWF)

In January 1988, the government of India approved the establishment of NCMRWF as a constituent unit of the Department of Science and Technology (DST) to help develop suitable numerical weather prediction (NWP) models for medium-range weather forecasts (3–10 days in advance) and prepare agrometeorological advisories for the farming community in 127 agroclimatic zones of India. The main objectives of NCMRWF are (1) to develop location specific medium-range (3 to 10 days) weather forecasts, (2) develop weather-based agro advisory services for the farming community, and (3) promote and coordinate research in related areas of meteorology and agrometeorology.

NCMRWF, in collaboration with the India Meteorological Department, Indian Council of Agricultural Research (ICAR), and state agricultural universities (SAUs), is providing Agrometeorological Advisory Service (AAS) at the scale of agroclimatic zones to the farming community, based on location-specific medium-range weather forecasts. To provide numerical weather prediction (NWP) based AAS to farmers, NCMRWF opened an agrometeorological field unit (AMFU) in each of the 127 agroclimatic zones of the country. These AMFUs are co-located with the National Agriculture Research Program (NARP) Centres of ICAR and the SAUs so that research output can be used effectively in formulating the agro advisories (www.ncmrwf.gov.in).

Central Research Institute for Dryland Agriculture (CRIDA): Data collected at different agrometerological observatories are used for drought assessment and their mitigations measures during the season by CRIDA. The weekly contingent crop plans for rainfed regions over the country during the crop growing seasons and at fortnightly interval in the rest of the period are prepared from the inputs provided by the centers of both the coordinating projects and the general weather situations prevailed over the period. These are updated regularly in the website "Crop-weather-outlook" maintained by CRIDA (www.crida.in).

Central Arid Zone Research Institute (CAZRI): CAZRI attempted classification of agricultural drought considering the values of Actual Evaporation/ Potential Evaporation (AE/PE) during different phenological phases of crop growth – seedling, vegetative and reproductive which was later improved with crop factor. CAZRI also maintains its own agrometeorological observatories in their regional research stations at Jodhpur, Jaisalmer, Chandan, Bikaner, Pali, and Bhopalgarh.

REMOTE SENSING USE FOR DROUGHT MONITORING AND EARLY WARNING

The remote sensing use for drought can benefit from climate variability predictions using coupled ocean/atmosphere models, survey of snow packs, persistent anomalous circulation patterns in the ocean and atmosphere, initial soil moisture and assimilation of remotely sensed data into numerical prediction models. The present state of remote sensing use for drought supports for all the phases of drought management activities which includes prediction, monitoring, early warning and drought relief management. Nearly-global seasonal climate anomaly predictions are possible due to the successful combination of observational satellite networks for operational meteorological, oceanographic and hydrological observations. Improved coupled models and near-real time evaluation of *in situ* and remote sensing data - allows for the first time physically-based drought warnings several months in advance, to which a growing number of countries already relate their policies in agriculture, fisheries and distribution of goods.

Satellite based remote sensing technology is widely used for monitoring crops and agricultural drought assessment. Over the last 20 years, coarse resolution satellite sensors are being used routinely to monitor vegetation and detect the impact of moisture stress on vegetation. The NOAA AVHRR NDVI has been extensively used for drought/vegetation monitoring, detection of drought and crop yield estimation (Baret and Guyot 1991, Batista et al 1997, Beneditte and Rossini 1993, Jeyaseelan et al 1987, 1990; Moulin et al 1998, NNRMS 1988, Tucker et al 1985). United States Agency for International Development has evolved a Famine Early Warning System NetWork (FEWSNET) by integrating the composite information on temperature, winds, humidity, soil and topography, observations on conflict, civil interest, health, market prices, field observations on agriculture, satellite derived rainfall and NDVI. FEWSNET is operationally issuing monthly food security reports for decision makers in Africa and USA (www.gsfc.nasa.gov). The Drought Monitor of USA using NOAA-AVHRR data (www.cpc.ncep.noaa.gov), Global Information and Early Warning System (GIEWS) and Advanced Real Time Environmental Monitoring Information System (ARTEMIS) of FAO using Meteosat derived cold cloud duration and SPOT – VGT derived decadal NDVI composites (Minamiguchi 2005), Inter-national Water Management Institute (IWMI)'s drought assessment in South west Asia using MODIS data (Thenkabail 2004) and NADAMS drought monitoring in India with IRS–WiFS/AWiFS and NOAA/AVHRR (Bhanuja kumari et al 2002, Jeyaseelan 1993, 2000, 2002, and 2003a,b, Jeyaseelan et al 1993, 1995, 1996, 1997, 1999 and 2001a,b, Jeyaseelan and Thiruvengadachari 1987a,b, Rao et al 2001, Thiruvengadachari 1988, Thiruvengadachari and Harikishan 1987, Thiruvengadachari et al 1991, 1993) data are the proven examples for successful application of satellite remote sensing for operational drought assessment

Remote sensing data provides wide ranging inputs for drought monitoring and early warning system. Time series data on different vegetation indices are extremely useful to detect the response patterns of agricultural crops to weather variations. Satellite meteorology provide inputs to all the three types of rainfall prediction systems currently in operation in India. In the long term rainfall prediction, using global and regional atmosphere, land and ocean parameters, the remote sensing data from geo-stationary and polar orbiting weather satellites, such as, INSAT, NOAA etc is being ingested directly or through parameterization. In the medium range weather prediction, the NCMRWF uses satellite based SST, NDVI, snow cover area and depth, surface temperature, altitude, roughness, soil moisture at surface level and TIROS Operational Vertical Sounder (TOVS) and Radio sonde data on water vapour, pressure and temperature at vertical profile data in the T86/NMC Model. However, at present only global data with poor spatial resolution is being used. In the short range rainfall prediction also INSAT based visible and thermal data is being used.

National Agricultural Drought Assessment and Monitoring System (NADAMS)

In India, National Agricultural Drought Assessment and Monitoring System (NADAMS) was initiated towards the end of 1986, with the participation of National Remote Sensing Agency, Dept. of Space, Government of India, as nodal agency for execution, with the support of India Meteorological Department (IMD) and various state departments of agriculture (Jeyaseelan et al 1997, Thiruvengadachari 1988, Thiruvengadachari et al 1991). NADAMS was made operational in 1990 and has been providing agricultural drought information in terms of prevalence, severity and persistence at state, district and sub-district level (NADAMS1990). Over a period of time, NADAMS project has undergone many methodological improvements such as use of moderate resolution data for disaggregated level assessment, use of multiple indices for drought assessment, augmentation of ground data bases, achieving synergy between ground observations and satellite based interpretation, providing user friendly information, enhanced frequency of information etc.

Monitoring of agricultural drought is limited to Kharif season (June-Oct/November), since this season is agriculturally more important and rainfall dependent. Resourcesat-1/Resourcesat-2 AWiFS based sub-district level agricultural drought assessment is carried-out in 4 states namely Andhra Pradesh, Karnataka, Maharashtra and Haryana. AVHRR/MODIS based district level assessment is carried out in 9 states namely Bihar, Chhattisgarh, Gujarat, Madhya Pradesh, Orissa, Rajasthan, Uttar Pradesh, Jharkhand and Tamil Nadu. Oceansat-2 OCM data was brought into operational use in kharif 2011 by generating fortnightly and monthly Rayleigh corrected TOA NDVI composites for all the 13 states. MODIS 250m derived NDVI and NDWI are also used for assessment. Coarse resolution products of soil moisture from passive microwave data, rainfall estimates and rainfall forecasts are also used in the assessment. The Area Favourable for Crop Sowing/crop sown area (AFCS), derived from Shortwave Angle Slope Index (SASI) and ancillary data is a recently added product in NADAMS project. This product is useful to monitor the sowing period drought conditions. Overview of NADAMS project is shown in Figure 2. From Kharif 2012, NADAMS project is being carried out by Mahalanobis National Crop Forecast Centre (MNCFC), Department of Agriculture and Cooperation, Ministry of Agriculture, with the technical support of NRSC (www.nrsc.gov.in).

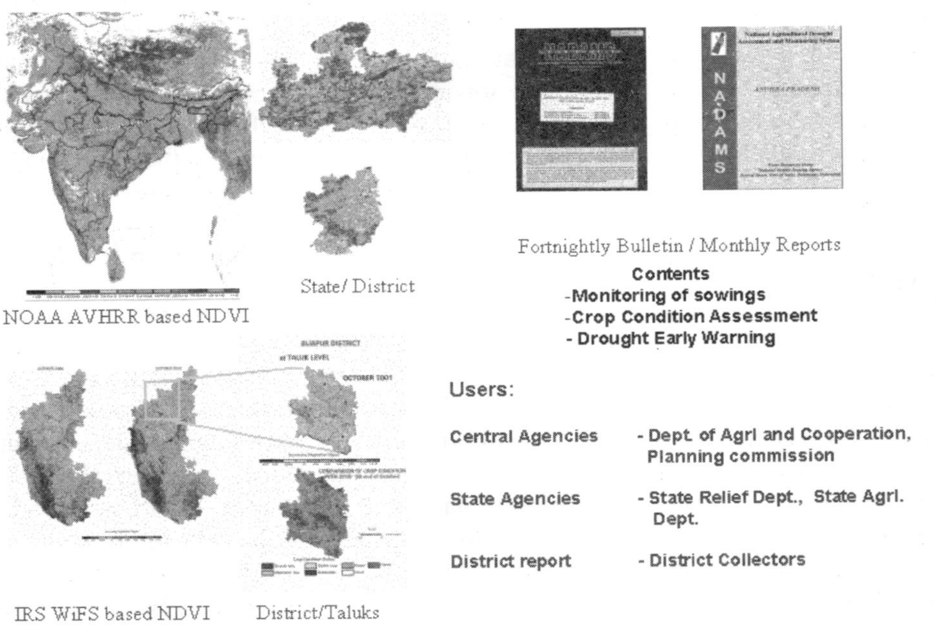

Fig. 2. Contribution by NADAMS

Forecasting Agricultural output using Space, Agrometeorology and Land based observations (FASAL)

Department of Space has initiated Cropped Acreage and Production Estimate (CAPE) project for major crops (*Kharif* rice, *Rabi* rice, wheat, jute, potato, mustard) at district level for the country with the support of various State Remote Sensing Applications Centres, State Departments of Agriculture and Agricultural Universities. Now FASAL is being carried out by Mahalanobis National Crop Forecast Centre (MNCFC) established at Dept. of Agriculture with the help of multi-institutional programme, which integrates the activity from many organisations such as Mahalanobis National Crop Forecast Centre (MNCFC), Space Applications Centre (ISRO), India Meteorological Department, Institute of Economic Growth and State Agricultural Departments. MNCFC has started providing crop forecasts, under FASAL programme, from the Kharif season of 2012.

Global Experience of Drought Monitoring and Early Warning

Drought monitoring mechanism exists in most of the countries based on ground based information on drought related parameters such as rainfall, weather, crop condition and water availability, etc. The data from Earth observation and Meteorological satellite are highly complementary to those collected by in-situ systems. Satellites are often necessary for the provision of synoptic, wide-area coverage and frequent information required to put in-situ information into spatial monitoring at regional, national and global scale. The present state of remote sensing use for drought monitoring and early warning is based on rainfall, surface wetness, temperature and vegetation monitoring is discussed

Currently, multi channel and multi sensor data sources from geostationary platforms such as GOES, METEOSAT, INSAT and GMS and polar orbiting satellites such as National Oceanic Atmospheric Administration (NOAA), EOS-Terra, Defence Meteorological Satellite Program (DMSP) and Indian Remote sensing Satellites (IRS) have been used for meteorological parameter evaluation, interpretation, validation and integration. These data are used to estimate precipitation intensity, amount, and coverage, measure moisture and winds, and to determine ground effects such as surface (soil) wetness.

The quality of seasonal predictions of temperature and precipitation anomalies by various centers such as the National Climate Research Centre (NCRC) of United States, European Centre for Medium Range Weather Forecasts (ECMWF), India Meteorological Department (IMD), National Centre for Medium Range Weather Forecast of India (NCMRWF), is also a function of the quality and amount of satellite data assimilated into the starting fields (e.g., SST from AVHRR and profiles from TOVS on NOAA satellites, ERS-2 scatterometer winds, SSM/I on DMSP satellites, all geostationary weather satellites: Geostationary Operational Environmental Satellites (GOES) i.e. GOES-East, GOES-West of USA, METeorological SATellite (METEOSAT) of Europe, Geostationary Meteorological Satellites (GMS) of Japan, Indian National Satellites (INSAT) of India etc. The new assimilation techniques have produced a stronger impact of space data on the quality of weather and seasonal climate predictions.

The potential contribution by existing satellites is by far not fully exploited, since neither the synergy gained by the combination of satellite sensors is used nor all the satellite data are distributed internationally. For example, better information flow from satellite data producers to the intermediary services like prediction centres such as CLIPS (Climate Information and Prediction Services) project of World Meteorological Organisation WMO, European Centre for Medium Range Weather Forecasts (ECMWF), National Centers for Environmental Predictions (NCEP), Japan Meteorological Agency (JMA), India Meteorological Department (IMD), National Centre for Medium Range Weather Forecast, India (NCMRWF) etc. to local services and ultimately to end users is not achieved. Further the drought predictions need to be improved with El Nino predictions and should be brought down to larger scales.

In case of drought monitoring and early warning based on normalised difference vegetation index (NDVI) and temperature condition index (TCI) derived from the satellite data is accepted world-wide for regional monitoring. The following regional/global monitoring and early warning programmes are ongoing at different countries/organisations. Africa Real-Time Environmental Monitoring using Imaging Satellites (ARTEMIS) is operational at FAO uses METEOSAT rainfall estimates and AVHRR NDVI values for Africa. The USDA/NOAA Joint Agricultural Weather Facility (JAWF) uses Global OLR anomaly maps, rainfall map, vegetation and temperature condition maps from GOES, METEOSAT, GMS and NOAA satellites for global monitoring. Joint Research Centre (JRC) of European Commission (EC) issues periodical bulletin on agricultural conditions under MARS-STAT (Application of Remote sensing to Agricultural statistics) project which uses vegetation index, thermal based evapotranspiration and microwave based indicators for European Countries. Agricultural Division of Statistics, Canada issues weekly crop condition reports based on NOAA AVHRR based NDVI along with agro meteorological statistics. National Remote sensing Agency, Department of Space/ Mahalanobis National Crop Forecast Centre (MNCFC), Dept of Agriculture, Government of India issues monthly reports at various administrative units under National Agricultural Drought Assessment and Monitoring System (NADAMS) which uses NOAA AVHRR and IRS WiFS based NDVI with ground based weather reports.

Three of the five new Earth science missions launched by NASA contribute to water cycle research and water-related national policy decisions.

Global Precipitation Measurement (GPM) is an international satellite mission of NASA and JAXA launched by a consortium of GPM partners in the United States, Japan, France, India, and Europe in 2014 to provide next-generation observations of rain and snow worldwide every three hours. The GPM missions helps our understanding of Earth's water and energy cycles, improve the forecasting of extreme events that cause natural disasters, and extend current capabilities of using satellite precipitation information to directly benefit society (http://www.nasa.gov/).

The ISS-RapidScat instrument launched to the International Space Station in 2014 monitored ocean winds to provide essential measurements used in weather predictions, including hurricane monitoring around the globe. The data are a key factor in climate research, weather and marine forecasting and tracking of storms and hurricanes (http://www.jpl.nasa.gov/).

The Soil Moisture Active Passive (SMAP), launched in January 2015 is an orbiting observatory that measures the amount of water in the top 5 cm of soil everywhere on Earth's surface. Soil moisture strongly affects plant growth and hence agricultural productivity, especially during conditions of water shortage and drought. Currently, there is no global in situ network for soil moisture monitoring. Global estimates of soil moisture and plant water stress must be derived from models. These model predictions (and hence drought monitoring) can be greatly enhanced through assimilation of space-based soil moisture observations. Improvements in the ability to monitor and forecast agricultural drought (i.e., the lack of root-zone soil moisture) will improve famine early warning in the most food-insecure countries in the world (http://smap.jpl.nasa.gov/).

NASA also plans to launch four additional water-related satellites in the next seven years: The Ice, Cloud, and land Elevation Satellite-2 (ICESat-2); Gravity Recovery and Climate Experiment (GRACE) Follow-on; Surface Water Ocean Topography mission; and the NASA-Indian Space Research Organisation Synthetic Aperture Radar mission (http://www.nasa.gov/).

REGIONAL DROUGHT MONITORING AND EARLY WARNING SYSTEMS

The United Nations Economic and Social Commission for Asia and the Pacific (ESCAP), the regional development arm of the United Nations for the Asia-Pacific region is implementing the Regional Cooperative Mechanism for Drought Monitoring and Early Warning. The Drought Mechanism, which is the flagship program of the Regional Space Applications Programme (RESAP) for Sustainable Development and the priority are highlighted by the Asia-Pacific Plan of Action for Applications of Space Technology and Geographic Information Systems for Disaster Risk Reduction and Sustainable Development, 2012-2017 (Asia-Pacific Plan of Action), in the overall context of building the agrarian economies of Asia and the Pacific region. The Mechanism envisages providing in-season satellite data, customized drought products and specialized capacity development training in order to enhance the knowledge and capacity of member countries for managing drought more efficiently. In 2013, ESCAP launched the Regional Drought Mechanism – a platform providing timely and free satellite-based data; products; and training to regional drought-prone countries – to enhance the capacity of Governments for agricultural monitoring and early warning. When combined with information collected on the ground, the data leads to more effective detection of potential drought conditions. As a Regional Service Node, National Remote Sensing Center, India Space Research Organization (ISRO) will provide not only the historical satellite images regarding drought in Sri Lanka in 2012-2013, but also the seasonal observation information as well as training for capacity development (www.escap.org.in).

CONCLUSION

Drought is one of the most devastating natural hazards in the world, claiming more lives and causing extensive damage to agriculture, vegetation, human and wild life and local economies. The good network of in situ observations on weather and climate parameters and its integration with the spatial synoptic observations from Earth Observation and Meteorological satellites available from Geostationary and Polar orbiting platforms significantly contributes in the activities of all the three major phases of drought management namely 1. Preparedness phase where activities such as prediction and risk zone identification are taken up long before the event occurs. 2. Prevention Phase where activities such as Early warning/Forecasting, monitoring and preparation of contingency plans are taken up just before or during the event and 3 Response/Mitigation phase where activities just after the event includes damage assessment and relief management. This article, briefly reviewed the drought monitoring and early warning system and the remote sensing use in a global scenario as well as in Indian Scenario.

REFERENCES

Baret F and Guyot G 1991 Potentials and limits of vegetation indices for LAI and APAR assessment, *Remote sensing of Environment* **35** 161-173

Batista TT, Shimabukuro YE, and Lawrence WT 1997 The long term monitoring of vegetation cover in the Amazonian region of northern Brazil using NOAA-AVHRR data, *International Journal of Remote Sensing* **18** 3195-3210

Benedetti R and Rossini P 1993 On the use of NDVI profiles as a tool for agricultural statistics: The case study of wheat yield estimate and forecast in Emilia Romagna, *Remote Sensing of Environment* **45** 311-326

Bhanuja Kumari V, Jeyaseelan AT, Prasad MA and Chandrasekar K 2002 Sub district drought assessment using IRS-WiFS data – Case study over Mahabubnagar District., ISPRS-TC-VII. In: International Symposium on Resource and Environmental Monitoring, Hyderabad

Blanford H F 1884 On the connexion of the Himalaya snowfall with dry winds and seasons of drought in India. In: Proceedings of Royal Society London **37** 1-23

Jeyaseelan AT 1993 Satellite based drought monitoring in India In: Workshop on application of remote sensing and GIS in Decentralised planning on drought and flood management Patna, 22nd June, 1993

Jeyaseelan AT 2000 Development of near real time drought management strategies, Doctoral Thesis, Center for Research, Jawaharlal Nehru Technological University, Hyderabad

Jeyaseelan AT 2002 Remote sensing applications to drought assessment and monitoring with case study on NADAMS In: National workshop on Recent trends in Drought assessment, monitoring and management, Center of Studies in Resources Engineering, Indian Institute of Technology, Mumbai

Jeyaseelan AT 2003(b) Remote sensing applications to drought management. In: National Seminar on GIS Application in Disaster Management held at National Institute of Rural Development, Rajendranagar, Hyderabad during August 18- 20, 2003

Jeyaseelan AT and Thiruvengadachari S 1987(a) Satellite monitoring of seasonal vegetation dynamics. Technical report No. RSAM-NRSA-DRM-TR-05/87, NRSA, Hyderabad, India

Jeyaseelan AT and Thiruvengadachari S 1987(b) Satellite monitoring of agricultural drought in Anantapur district of Andhra Pradesh, Technical report No. RSAM-NRSA-DRM-TR-10/87, NRSA, Hyderabad, India

Jeyaseelan AT 2003(a) Droughts & floods assessment and monitoring using remote sensing and GIS. In: International training Workshop on Satellite Remote sensing & GIS Applications in Agricultural Meteorology for WMO member countries in Asia Region, 7-11 July,2003, CSSTEAP/IIRS, Dehradun

Jeyaseelan AT, Chandrasekar K and Suresh babu AV 2001 Satellite based drought monitoring and Cyclone damage assessment in Orissa during 1999. Symposium on the impact of human activities on Thar Desert Environment, Jodhpur, Feb 15-17 2001, India

Jeyaseelan AT, Chandrasekar K and Thiruvengadachari S 1997 Satellite (NOAA) based monitoring of Bhakra irrigation Command, ISPRS. In: Workshop on Application of Remote sensing and GIS for Sustainable development, 24-25 November 1997, Hyderabad, India

Jeyaseelan AT, Giridhar MVSS and Thiruvengadachari S 1995 Validation of satellite derived crop stress assessment through water balance approach at mandal level'. In: National Symposium of Advanced Technologies in Meteorology, TROPMET - 95, 8-1 February 1995, NRSA, Hyderabad

Jeyaseelan AT, Sujatha CL, Thiruvengadachari S 1996 Drought Monitoring in India - A Case Study over Andhra Pradesh. In: International conference on Disaster and Mitigation, Anna University, Madras, 19-22, January 1996

Jeyaseelan AT, Suresh babu AV, Chandrasekar K and Rupesh kumar GV 2001(b) IRS WiFS data use for 1999 droughts in Rayalaseema districts of Andhra Pradesh, India. In: Proceedings of International Conference on Remote Sensing and GIS/GPS, (ICORG-2000), Hyderabad, India

Jeyaseelan AT, Sureshbabu AV, Chandrasekar K 1999 Improvement in Agricultural Drought Monitoring using IRS WiFS in Andhra Pradesh State during 1998. In: Proceedings of ISRS National Symposium on Remote sensing Applications for Natural Resources, Retrospective and Perspective, January 19-21, 1999, Bangalore

Jeyaseelan AT, Vidya S and Thiruvengadachari S 1993 GIS based agricultural drought impact characterisation - Case Study over Mahaboobnagar District. In: International Seminar on Disasters Environment and Development by IDNDR, New Delhi

Minamiguchi N 2005 The Application of geospatial and disaster information for food insecurity and agricultural drought monitoring and assessment by the FAO GIEWS and Asia FIVIMS. In : Workshop on Reducing Food Insecurity Associated with Natural Disasters in Asia and the Pacific. Bangkok, Thailand, 27-28 January 2005. FAO Regional Office for Asia and the Pacific, Bangkok, Thailand

Moulin SA, Bondeau A and Delecolle R 1998 Combining agricultural crop models and satellite observations: From field to regional scales, *International Journal of Remote Sensing* **19** 1021-1036

NADAMS 1990 Seasonal conditions during June –December 1989 and user feedback, NADAMS, NRSA, Hyderabad, India

NNRMS 1988 Operations manual on vegetation index mapping using NOAA data for drought mission, National Natural Resources Management Systems, Technical report no: RC-DRM-88/01

NOAA AVHRR - A case study over Karnataka. In: All India Seminar on Natural Disaster - Causes and Management, Dec 7-9 DMCT, Bangalore

Rao DP, Diwedi RS and Jeyaseelan AT 2001 Role of Space technology in the assessment and monitoring of Drought and Desertification. In: Symposium on the impact of human activities on Thar Desert Environment, Jodhpur, Feb 15-17 2001, India.

Sastri ASRAS 1993 Agricultural Drought Management Strategies to Alleviate Impacts: Examples from the arid and sub humid regions of Indian subcontinent, In: Drought assessment, management and planning: Theory and case study Ed. Donald A Wilhite, Springer Science, 65-86 pages

Thenkabail P S 2004 Inter-sensor relationships between IKONOS and Landsat-7 ETM+ NDVI data in three eco regions of Africa, *International Journal of Remote Sensing* **25**(2) 389 – 408

Thiruvengadachari, S and Harikishan, J 1987. Satellite Monitoring of Agricultural Droughts in Hindupur Division in Anantapur District of Andhra Pradesh State. Technical Report No. RSAM-NRSA-DRM-TR-08/87, NRSA, Hyderabad, India.

Thiruvengadachari S 1988 National agricultural drought assessment and monitoring system in India: The satellite contribution. In: 39th IAF congress, special publication on Space and Drought Management, ISRO-44-89, Bangalore, India

Thiruvengadachari S, Jeyaseelan AT, Harikishan J and Krishna Rao PV 1991 Satellite surveillance of agricultural drought conditions in India. Published in the book on *Application of remote sensing in asia and oceania - environment change monitoring* edited by Sunjimurai

Thiruvengadachari S, Jeyaseelan AT, Murthy C S and Shanker M 1993 Satellite based crop condition assessment over India during 1993 *Kharif*. In: 30th Annual convention of Indian Geophysical union and Seminar on Space Applications to Earth System Science, NGRI, Hyderabad 21 - 23 Dec 1993

Tucker CJ, Townshend JRG and Goff TE 1985 African land-cover classification using satellite data, *Science* **227** 369-375.

4

Weather Forecast and Agromet Advisory Services for Drought Management

D.V. Singh

INTRODUCTION

Rainfall during the monsoon season accounts for about 75-80% of the total annual rainfall in India. Entire country except for southern peninsula, Jammu & Kashmir and Assam receives more than 75% of its annual rainfall during this season. In Saurashtra and Kutchh in Gujarat and adjoining Rajasthan and Madhya Pradesh, the percentage is more than 90%. The monsoon rainfall has significant effect on crop production during *kharif* season and to some extent during following *rabi* season.

The most significant feature of the annual cycle of Indian summer monsoon is its regularity. The main components of southwest monsoon over India are Mascarene high, Somali low level jet, Tibetan anticyclone and tropical easterly jet. The location and intensity of these components govern the strength and spatial distribution of rainfall over India. Also, the semi-permanent features of southwest monsoon viz. monsoon trough with embedded vortices control the rainfall activity. The major weather systems contributing to the rainfall during the season are: monsoon lows and depressions, off-shore trough and vortices and mid-tropospheric cyclonic circulations.

Deficit rainfall, relative to the long-term average, for an extended period of time, usually during a season when substantial rainfall is normally expected over the area results in drought. It is a normal, recurrent feature of climate and is observed in all the climatic zone. The severity of the drought can also be aggravated by other climatic factors such as high temperature, high wind speed and low humidity. Drought is still largely unpredictable and varies with regard to

---

Senior Scientist, Central Arid Zone Research Institute, Jodhpur- 342 003
Email:dvsingh.cazri@gmail.com

the time of occurrence, duration, intensity and extent of the area affected. At national level, a year is considered as a 'drought year' when the area affected by moderate and severe droughts either individually or together is 20-40% of the total area of the country and seasonal rainfall deficiency during southwest monsoon season for the country as a whole is at least 10% or more. If the spatial coverage of drought is more than 40%, it is known as all India severe drought year. Based on the per cent departure of rainfall from normal during 1875-2009, western Rajasthan (34 years) and Saurashtra & Kutch (31 years) regions of Gujarat were found to be the most drought prone areas.

Thus, there is a pressing need to understand the Indian monsoon and forecast its inter-annual and seasonal variability. Forecasting means the process of estimation of the value of some variable at some future time. Forecast of weather parameters such as rainfall, temperature, wind, humidity etc. over a region averaged over a particular time period is one of the primary functions of the national weather services. The Agro advisory services of the nation collect and organise climate/weather, soil and crop information and amalgamate them with weather forecast to assist farmers in taking management decisions. The seasonal climate forecasts before the start of cropping season would help the farmer to adapt agriculture system to ensuing weather variability. Subsequent to this, short and medium range weather forecast based agrometeorological advisories become vital to stabilize their yields through management of agroclimatic resources as well as other inputs such as irrigation, fertilizers and other pesticides.

TYPES OF WEATHER FORECASTS

Depending on the spatial and temporal scales of atmospheric systems and the details of the accuracy desired, the weather forecasts are divided into following categories.

Now Casting: The details about the current weather and forecasts up to a few hours (< 24 hours) ahead are given. Now casting may include forecast of many weather variables including wind, temperature, precipitation and fog.

Short Range Forecast (SRF): The weather in each successive 24 hour intervals may be predicted up to 72 hours. This forecast is mainly based on the weather systems observed in the latest weather charts, although generation of new systems is also considered.

Medium Range Forecast (MRF): Weather forecast is for 3 to 10 days. Average weather conditions and the weather on each day may be predicted. The forecast is usually based on Numerical Weather Prediction (NWP) products.

Long Range /Extended Range Forecasts (LRF/ERF): There is no rigid time frame of LRF, which may range from 10 days to a seasonal forecast. As per the World Meteorological Organization (WMO), long range forecast is defined as the forecast from 30 days' up to one season's description of average weather parameters. The monthly and seasonal forecast comes under long range forecast.

Weather forecast system in India

India Meteorological Department (IMD) is solely responsible for issuing operational forecast for India. In view of growing operational requirements from various user agencies, IMD has embarked on a seamless forecasting system covering short range to extended range and long range forecasts. Though IMD is the only agency mandated for providing weather forecasts, many institutes in India are involved in the research work related to forecast. Some of these institutes are Indian Institute of Tropical Meteorology (IITM), Pune, Indian Institute of Science (IISc), Bangalore, Space Applications Centre (SAC), Ahmedabad, National Aerospace Laboratories (NAL), Bangalore, Centre for Mathematical Modeling and Computer Simulation (CMMACS), Bangalore, National Centre for Medium Range Weather Forecasting (NCMRWF), Noida and Centre for Development of Advanced Computing (C-DAC), Pune. Many international climate centres like National Centre for Environmental Prediction (NCEP), European Centre for Medium-Range Weather Forecasts (ECMWF), International Research Institute for Climate and Society (IRI) etc. are also involved in the research related to seasonal prediction of monsoon rainfall as a part of their efforts to improve global forecasts. IMD makes use of experimental forecasts prepared by these climate research centres both inside and outside India as supportive materials for preparing the operational forecasts for India.

There are various methods that are used to create weather forecasts of different durations. In India, LRF are based on empirical (statistical) methods, while MRF are usually based on Numerical Weather Prediction (NWP).

Long Range Forecasts (LRF)

The LRF provide a range of possible changes that are likely to occur in the season ahead. However, it is not possible to predict the daily weather variations at a specific location months in advance because of the chaotic nature of the atmospheric circulation. The long range forecasts are possible to some degree due to a number of components which though show variations on long time scales (seasons and years) are to a certain extent predictable (ECMWF 2005). The most important of these components is the ENSO (El Nino Southern Oscillation) cycle that refers to the coherent, large-scale fluctuation of ocean temperatures, rainfall, atmospheric circulation, vertical motion and air pressure across the tropical Pacific. It is the largest known source of year-to-year climate variability (ECMWF 2005). Unusually warm or cold sea surface temperatures in the tropical Atlantic or Indian Ocean can also cause major shifts in seasonal climate in nearby continents. In addition to the tropical oceans, other factors that may influence seasonal climate are snow cover and soil wetness. When snow cover is above average for a given season and region, it has a greater cooling influence on the air than usual. Soil wetness also has a cooling influence. All such factors affecting the atmospheric circulation constitute the basis of long-term predictions.

The first regular forecast of Indian Summer Monsoon Rainfall (ISMR) was issued by the India Meteorological Department (IMD) in 1886. Since then, ISMR forecast is continuing till date but with several changes in its format and content. From 1886, the monsoon forecasts were issued for the entire India and Burma. In 1922, India was classed into three main homogenous areas, namely, (i) Peninsula (ii) Northeast India and (iii) Northwest India. In 1935, forecast for NE India was discontinued for wants of suitable predictors and skill of model. The practice of issuing forecasts for two homogenous regions of India (northwest India and Peninsula) was continued till 1987.

IMD introduced the 16 parameter power regression and parametric models in 1988 and started issuing forecasts for the southwest monsoon rainfall over the country as a whole. Quantitative forecasts were issued using the power regression model, and qualitative forecasts (normal/excess or deficient) were issued using the parametric model. After the failure of forecast in 2002, IMD introduced a new two stage forecast strategy in 2003. Accordingly, the first stage forecast for the seasonal (June to September) rainfall over the country as a whole was issued in April and the update for the April forecasts was issued in June. Along with the updated forecast, forecast for seasonal rainfall over broad homogeneous rainfall regions of India and July rainfall over country as a whole were also issued. During 2003-2006, the first stage quantitative and 5 category probabilistic forecast for the season rainfall over the country as a whole were issued using 8-parameter power regression (PR) model and linear discriminant analysis (LDA) model, respectively. Update for the first stage forecasts were issued using 10-parameter PR and LDA models. In 2004, the country was reclassified into 4 sub geographical regions. In 2007, IMD introduced new statistical forecasting system based on ensemble technique for the South-West monsoon (June - September) rainfall forecast. In 2009, forecast for August rainfall over country as a whole was started along with other second stage forecasts issued in June.

At present, the forecast for the ISMR rainfall is issued in two stages. The first stage forecast for the seasonal (June to September) rainfall over the country as a whole is issued in April and the update of the April forecast is issued in June. Five predictors are used for April forecast and for the updated forecast in June, six predictors are used (Table 1). Along with the updated forecast, forecast for seasonal rainfall over four broad geographical regions of India and July rainfall over country as a whole are also issued. From 2010, operational forecast for the rainfall during the second half of the monsoon season (August-September) and that during the September over the country as a whole are also issued.

Table 1. Predictors considered for forecast of SW monsoon rainfall in India

S No	Predictor (Period)	Used for the forecasts in
1	North Atlantic Sea Surface Temperature (December + January)	April and June
2	Equatorial SE Indian Ocean Sea Surface Temperature (February + March)	April and June
3	East Asia Mean Sea Level Pressure (February + March)	April and June
4	NW Europe Land Surface Air Temperatures (January)	April
5	Equatorial Pacific Warm Water Volume (February + March)	April
6	Central Pacific (Nino 3.4) Sea Surface Temperature Tendency (MAM-DJF)	June
7	North Atlantic Mean Sea Level Pressure (May)	June
8	North Central Pacific Wind at 1.5 Km above sea level (May)	June

IMD issues forecasts for monthly (for July, August, September), second half (August + September) and seasonal rainfall over the country as a whole and for seasonal rainfall over four geographical regions (Northwest India, Central India, Northeast India and South Peninsula) with useful skill.

In the ensemble technique, all possible models based on all the combination of predictors are developed. For model development, two different statistical techniques namely, Multiple Regression (MR) and Projection Pursuit Regression (PPR) are considered. Out of all the possible models, best few models are selected based on their skill in predicting monsoon rainfall during a common period. Ensemble mean is computed as the weighted average of the forecast from the selected models. While the MR technique is a conventional linear model technique, the PPR method is known for its superiority in capturing the non-linear relationships between the predictors and rainfall. Verification of the results with the past data showed that the ensemble method performed better than the individual models.

For forecasting of South-West monsoon season rainfall over the four broad geographical regions of India (northwest India, central India, south peninsula and northeast India), multiple regression (MR) models based on separate set of predictors are used.

IMD also prepares operational long range forecasts for winter precipitation (January to March) over northwest India and northeast monsoon rainfall (October to December) over south peninsula. For this purpose, separate statistical models have been developed.

Under National Mission for development of dynamical models for the monsoon forecasting, in near future IMD is planning to setup a dynamical coupled model prediction system for monthly and seasonal prediction with the help of IITM.

Medium Range Forecast (MRF)

MRF are based on the output of one or more numerical weather prediction (NWP) models. Complex computer programs, also known as forecast models, are run on supercomputers that provide predictions on many atmospheric variables such as temperature, pressure, wind and rainfall. NWP models solve a complex set of hydrodynamic equations that describe the evolution of the atmosphere, subject to the initial atmospheric state. The NWP method is flawed in that the equations used by the models to simulate the atmosphere are not precise. This leads to some error in the predictions. In addition, there are many gaps in the data of initial condition. If the initial state is not completely known, the model's prediction of how that initial state will evolve will not be entirely accurate. Due to these limits the skill to forecast the weather deterministically deteriorate beyond 6 days or so, depending on the season. Despite these flaws, the NWP method is probably the best method of forecasting the day-to-day weather changes in short and medium range.

To improve the model skill, instead of using just one model run, many runs with slightly different initial conditions are made. An average, or 'ensemble mean', of the different forecasts is created. This ensemble mean is likely to be better because it averages over the many possible initial states and essentially smoothens the chaotic nature of climate. In addition, it is now possible to forecast probabilities of different conditions because of the large ensemble of forecasts available. Very recently, IMD has implemented a multi-model ensemble (MME) based five days quantitative forecasts in medium range at district level.

AGRO METEOROLOGICAL ADVISORY SERVICES IN INDIA

The advent of new weather forecasting capabilities provides opportunities for the meteorological community to develop better products and information to decision-makers in the various user sectors (i.e. agriculture, energy, health). Some examples of using NWP related to the environmental conditions that influence agricultural production include monsoon systems (Mishra and Krishnamurti 2007) and heavy rains (Fall et al 2007). Both LRF and NWP forecasts can be used in agricultural decision making. Tactical decisions like sowing, frost protection, irrigation, fertilization disease and pest management, etc. can benefit from the utilization of weather predictions. The value of meteorological forecasts is expected to increase as the agricultural system becomes more susceptible to climate variability under the effects of climate change.

World Meteorological Organisation has also emphasized that short- to medium-range weather forecasts and seasonal outlooks represent a key service for agriculture. Users want to use the output from various forecast models for their specific applications and decision-making processes. The use of meteorological forecast products in agricultural decision making problems has a long history. Direct NWP output has been used in predictions of irrigation, soil moisture, days of fieldwork, timing crop harvests and crop / animal pest and disease development.

In Indian context, the working group (WG) of Planning Commission of India on agriculture research and education for the eleventh five year plan (2007-2012) emphasized that there is an urgent need to review the strategies of agricultural research programs in relation to our understanding of the climate phenomena responsible for variable production and for sustaining food production and forward planning. The WG also stressed that in order to implement the programs for efficient natural resource management, it is necessary to address agrometeorological issues for weather-wise agricultural management. The WG recommended that the existing agro-advisory systems need to be value-added and developed for application as an operational tool, so that weather related uncertainties are managed wisely.

Agrometeorological or agromet Advisory is advice by experts to crop growers/livestock producers on the basis of possible future weather and climate conditions on 'what to do' or 'what not to do' for maximizing the advantages and minimizing the losses in production. Weather forecasts have little importance unless they are operationally used.

For maximum advantage of weather forecasts, agromet advisories are issued in consultation with experts of various disciplines considering the past, present and predicted weather and its spatial-temporal behaviour *vis-a-vis* crop condition. Appropriate weather forecast has tremendous benefits in terms of pre-facto management of the negative impacts of vagaries of weather. This is because the cost of pre-facto risk reduction due to weather is much smaller than the post-facto management of the losses (Rathore et al 2006).

Das et al (2010) opined that short and medium range forecasts are important for farmers to plan the farm operations such as:

- Preparatory activities, such as land preparation and preparation of plant material
- Planting or seeding/sowing
- Crops, fruit trees and vine management; application of fertilizer, irrigation; thinning, topping, weeding; pest and disease control
- Management of grazing systems
- Harvesting, on-farm post-harvest processing and transport of produce
- Livestock production (dairy enterprises, beef systems, lamb and other livestock systems)

Broadly, the agromet advisories recommend

i. Implementation of certain practices or the use of special materials to help effectively prevent or minimize possible weather-related crop damage or loss.

ii. Initiation of cultural practices which are weather sensitive.

Furthermore, quantitative forecasts are an important source of data for simulation models that produce information useful for farmers (simulation of crop phenology, water and nutrient cycles, crop production, weed, disease and pest cycles, etc.). The model outputs may then be used in preparation of agromet advisories.

An added advantage of such services is that they help in reducing the environmental pollution through the optimal use of agricultural chemicals. Agromet advisories are being issued by almost all the developed and developing countries on various spatial and temporal scales.

In India, Agromet Advisory Services (AAS) are being implemented successfully through collective efforts of IMD, State Agricultural Universities, Indian Council of Agricultural Research (ICAR) institutes, IITs, NGOs and private partners. Earlier the project was known as Integrated Agromet Advisory Service (IAAS) and the project name was changed to Gramin Krishi Mausham Sewa (GKMS) in 2013. From the inception of the project, commendable services are being provided to the farmers which in turn increase the productivity of crops and ultimately improve the economic condition of the low and marginal farmers in the country. A number of important activities like preparation of district level weather forecast and agromet advisories, multi-channel dissemination of advisories, popularization and awareness programme, human resource development, R & D, installation of observatories and its maintenance, etc. are being carried out in this project. Phenomenal achievement in dissemination of advisories under Public Private Partnership (PPP) mode is achieved. Involvement of farmers with this service through organization of Farmer Awareness Programmes (FAP) is encouraged.

Preparation of Agromet Advisories

The AAS is provided to the country's farming community in the form of agromet bulletins. The advisories are prepared by the experts and are tailored to the requirements of farmers of the district.

The preparation of agromet advisories requires close links among various data-providers and experts from different fields. The basic requirement is that the forecasted data must be for the desired period and for the specific location under consideration.

IMD provides district level weather forecast to AgroMet Field Units (AMFUs), located at State Agricultural Universities, ICAR institutes and IITs in different parts of country, twice a week on every Tuesday and Friday. This MRF is based on multi-model ensemble of NWP models. The quantitative weather forecast includes daily values of eight weather parameters, viz. rainfall, maximum and minimum temperature, cloud cover, maximum and minimum relative humidity, wind speed and wind direction for coming five days.

The AAS bulletins are issued at district, state and national level. The district level AAS bulletins are issued by Agromet Field Units (AMFUs). These bulletins include crop specific advisories for field crops, horticultural crops and livestock. At present these bulletins are issued for 563 districts of the country. A panel of experts discuss the present, past and future status of weather and current crop conditions and recommends the appropriate operations for better farm management based on such forecasts. Priority is given to predominant crops of the region and most prevailing problems keeping in view the relative economic importance.

Management practices like what, when and how to sow; when and how much to irrigate; what measures to be adopted for plant and animal protection from stresses caused by pest and disease, temperature, wind, and rainfall, etc. are suggested. Animal shelter, nutrition and their health are affected by weather to a large extent and hence are given a place in these services. On the basis of local agrometeorological and farming information and the weather forecasts, the subject matter specialists discuss the options and consequent effects and then decide the advisories and options for the action by the farmers in respect of the items related to their expertise (Singh et al 1999).

State Level AAS bulletin is a composite of district AAS bulletins. These bulletins are jointly prepared by Meteorological Centres of IMD, State Agricultural Department and AMFUs and mainly used by functionaries of the concerned state. These are also useful to fertilizer industry, pesticide industry, irrigation department, seed companies, transport and other organizations dealing with inputs in agriculture. The bulletins are significant input to the state level Crop Weather Watch Group (CWWG) meeting. Presently, these bulletins are issued for all the states in the country.

National level AAS bulletin is prepared by National Agromet Advisory Service Centre, Division of Agriculture Meteorology, IMD, Pune, with inputs from various State Meteorological Centres and AMFUs. This bulletin helps identify stress on various crops for different regions of the country and suitably incorporate advisories. National agromet advisory bulletin is primarily targeted for national level planners e.g. Crop Weather Watch Group (CWWG), Department of Agriculture and Co-operation (DoAC) and also communicated to all the related ministries (state and central), several organizations and NGOs for their use.

Impact Assessment

An Impact assessment of AASs was guided and monitored by national committee of experts in India where 40 AAS and 40 non-AAS farmers were surveyed. These farmers were randomly sampled and were generally middle-aged and had medium-to-large land holdings. The data revealed that inputs used varied qualitatively and significantly between AAS and non-AAS farmers. AAS farmers received agro-advisories based on medium-range weather forecasts including optimum use of inputs for different farm operations. The timeliness of proper agro-advisories given for various farm operations might have saved the crops from possible moisture stress, nutritional stress and pest attack. Non-AAS farmers used almost the same quality of inputs, but their timing of applications was different from that of the AAS farmers resulting in substantial increase in profit (Table 2).

Table 2. Economic impact of AAS based on medium-range weather forecast

Crop	Station	% change in cost of prod. (per ha)	% change in crop yield (per ha)	% change in profit (per ha)
Cotton	Hissar	1	14	10
	Coimbatore	- 4	16	16
Rice	Ludhiana	- 6	9	18
	Kalyani	- 3	21	29
Wheat	Ludhiana	- 6	9	17
Mustard	Hissar	- 3	8	13

Recent initiatives for dissemination of AAS bulletins

The AAS bulletins are disseminated through newspapers, TV, Radio, Internet, SMS, personal contacts, etc. AAS has so far partnered with Farmers Portal, Reuter Market Light (RML), Handygo, State Department of Agriculture (Government of Maharashtra), IFFCO Kisan Sanchar Ltd. (IKSL) and NOKIA Life Tools for dissemination of brief advisories through SMS. In the XIIth Five Year Plan, it is proposed to include more number of public/ private agencies/ companies for the dissemination of agromet advisories so that the agromet SMS can reach to more farmers.

Collaboration with additional new private companies is envisaged for further outreach. One of them is Reliance Foundation that has embarked upon a new initiative called Reliance Knowledge for Sustainable Development (RK4SD). RK4SD aims at capacity building of farmers, fisher folk and micro-enterprises for better livelihood security. RK4SD provides timely advisories on climatic conditions, water and weather related risks, methods of controlling pests and diseases and sharing of better farming methods and best management practices. Another private company is 'Infronics Solutions' that has developed 'Devolved Decision Making System' which is beneficial to government, farmers and agribusiness firms. Zoraly Solutions of Infornics is working in agribusiness services like farming, dairy, post-harvest, supply chain management and marketing for the farmers. IMD and Infronics will be collaborating to communicate agricultural information to the farming community mainly on weather forecasting and advisories as per unique needs of farming community.

In addition to these, the other proposed projects for dissemination of agromet advisories are Tata Consultancy Services (TCS), National Bank for Agriculture and Rural Development (NABARD), Common Service Centre (CSC), ITC e-Choupal, M.S. Swaminathan Research Foundation, IBM, Infosys, Brand TV, Monsanto and Centre for Agricultural Bioscience International (CABI).

Future plan for improvement of AAS

- Research and development component needs to be enhanced to improve the quality of agromet advisories.
- Strengthening of state level activities like review and capacity building.
- Improve the weather forecast through value addition at state level.
- Linkages with line departments at state level needs to be strengthened to improve the GKMS activities. Linkages will lead to better information about crop condition, out-reach and feedback, awareness among the users.
- Block-level database for weather and climate, crops, contact details of farmers and other related information for developing block level agromet advisories need to be created.
- Block level advisories will be started at selected KVKs through establishment of District Agromet Units during first phase.
- The AMFUs will start using e-Agromet software for preparation of agromet advisories.
- The AMFUs will identify new sectors and issue agromet advisories for these sectors.
- Dissemination of agromet advisories through Farmers portal needs to be increased.
- The AMFUs will take initiative to popularize GKMS through Farmers Awareness Programme, participation in farmers' fairs, field demonstrations, etc.
- Advisories on animal husbandry and inland fisheries need to be strengthened.

REFERENCES

Das HP, Doblas–Reyes FJ, Garcia A, Hansen J, Mariani L., Nain A., Ramesh K, Rathore LS and Venkataraman R 2010 Weather and Climate Forecasts for Agriculture. *In Guide to Agricultural Meteorological Practices (GAMP)*. Geneva, CH: WMO.

ECMWF 2005 The basis of seasonal forecasting. (http://www.ecmwf.int/products/forecasts/seasonal/)

Fall S, Niyogi D, Mohanty UC and Kumar A 2007 Application of weather prediction models for hazard mitigation planning: a case study of heavy off-season rains in Senegal. *Natural Hazards* **41(1)** 227-243

Mishra AK and Krishnamurti TN 2007 Current status of multimodel superensemble and operational NWP forecast of the Indian summer monsoon. *Journal of Earth System Science* **116(5)** 369-384

Rathore LS, Parvinder M and Kaushik S 2006 Impact assessment of the agro-meteorological advisory service of the National Centre for Medium Range Weather Forecast (NCMRWF). *Available on the INSAM website (www.agrometeorology.org) under 'Accounts of operational agrometeorology'*.

Singh SV, Rathore LS, Trivedi HKN 1999 Guide for Agrometeorological Advisory Service. National Centre for Medium Range Weather Forecasting, Department of Science and Technology, Government of India, New Delhi-110003, India.

5

Use of Remote Sensing in Drought Assessment and Monitoring in Western Rajasthan

P.C. Moharana

INTRODUCTION

Drought can be defined as a recurring extreme climatic event over land, characterized by below normal rainfall over a period of months to years (Dai 2011). The common perception about drought is that it is a regional weather phenomenon that occurs in all climatic zones. However, its intensity and frequency usually increases towards drier zones. In a country like India, where agriculture is the major source of livelihood and that by and large depends on the monsoon, drought often leads to shortage of food, fodder and water. Considering that the arid regions receive much less rainfall than the other regions on the earth and also less number of rainy days, the incidence of drought is also higher in this region. In Indian part of arid zone, which is more confined to western part of Rajasthan in 62% of country's arid lands, drought is a frequent phenomenon. Between 1901 and 2010, the region experienced droughts of moderate to severe intensities in last 57 years. In recent decades, the drought of 2002 was of disastrous proportion when annual rainfall was mere 127 mm (38% of the long-term average) and the crucial summer monsoon rainfall was 90 mm, against the normal of 280 mm (Saha et al 2012).

Drought is not easy to predict well in advance for most locations for simple reason that it depends how precisely we predict two variables, precipitation and temperature. One explanation for this is that droughts are a slow-onset, or "creeping," natural disaster, developing over months and years, and frequently they exist before it is realized that a drought is occurring (Albert et al 2002). Secondly, a drought's severity varies by its precipitation deficit, spatial extent,

Principal Scientist, Central Arid Zone Research Institute, Jodhpur-342003
E-mail: pcmoharana45@gmail.com

and duration, making it difficult to compare one drought to another. Finally, the impacts of drought are based on the range of economic, environmental, and social resources within a region.

Table 1. More technical definition on types of drought is mentioned in the following table.

Type of droughts	Definition	Database requirement
Meteorological drought	It is the situation when actual rainfall is less than 25% or more than the normal rainfall over a wide area. As per IMD (Indian Meteorological Department), rainfall distribution is considered "normal" when its deviation from the long term average in a particular year is between 0 and 19%. Between 20 and 59 % is called a deficit and 60 or more is termed as scanty.	Daily rainfall, temperature, humidity, wind velocity, pressure, evaporation
Hydrological drought	This is an extension of meterological drought deciphered from lake and reservoir levels, ground water level. Hydrological drought takes a longer time than agricultural or meteorological droughts to manifest.	Surface area & volume, run-off, river flow measurement, infiltration for surface water Water level fluctuation, aquifer characteristics for sub-surface water availability
Agricultural drought	It occurs when soil moisture and rainfall are inadequate during the growing season to support healthy crop growth to maturity and cause extreme crop stress. IMD calculates agricultural drought in terms of aridity index measured as (AE-PE) x 100/PE) AE, mean evapo-transpiration, PE, potential evapo-transpiration. It is calculated on weekly basis. < 26 % = mild, 26 –50% = moderate, >50% = severe	Soil texture, soil fertility, soil moisture, crop type and area and crop water requirement

Drought is perceived different ways by different people. Its definition surrounds around specific objectives and type of drought. For example, to a hydrologist it is a condition of less water availability in water bodies like streams, wells, tanks and reservoir, for an agronomist : soil moisture deficiency, crop yield reduction and for a meteorologist, it is about occurrence of less rainfall. However, to a common man, it is the scarcity of essential commodity (mainly food crops), drinking water, rise in price and unemployment. These are otherwise known as operational definitions (Table 1) of drought.

Importance of drought assessment: In India, more than 70 percent population depend on agriculture, 68% of net sown area (142.2 M ha) is drought prone and 50% drought prone is severe in nature (NRSC, 2010). At this scale, India experiences localized drought almost every year in some region or other. Major droughts that affected our country in post-Independence were reported during 1951, 1966-67, 1972, 1979, 1987-88 and 2002-03 (Subbaiha, 2004). Thus despite advances in technological advances, Indian agriculture continues to be affected by droughts. Arid and semi-arid regions of India are more prone to drought than the other climatic zones with a probability of drought being >20%. Areas where more than 40% of the rainfall recording years have experienced drought are classified as chronically drought-prone while those experiencing drought for 20-40% years are termed as drought prone (CAZRI, 2005). The hot arid regions of our country (32.7 m ha) in western Rajasthan, Gujarat, Haryana and part of Karnataka and A.P. are chronically drought prone.

Therefore, irrespective of scale and severity of problem, it is always an admissible fact that, occurrences of drought leads to a chain of impacts; crop failure-cattle starvation-unemployment and migration-fodder shortage, and therefore, larger impact on the state's economy. It is at this juncture, management of drought becomes a prime priority which needs information. Geo-spatial information based assessment therefore, becomes a preferred tool.

REMOTE SENSING TECHNIQUES USED FOR DROUGHT INFORMATION

Satellite sensors provide spatial information on vegetation stress caused as an impact of drought conditions which is a useful information for the Government sector to understand and address the problem. Satellite remote sensing technology is widely used for monitoring crops and drought assessment. Before we look at those sensors, let us discuss some of the characteristics of remote sensing technology that needs to be understood for its application in drought assessment.

Electromagnetic Spectrum: Electromagnetic spectrum (EMS) represents the continuum of electromagnetic radiation (EMR) arranged on the basis of wavelengths or frequency. This spectrum ranges from shorter wavelengths (gamma rays to X rays) to the longer wavelengths (microwave and radio waves). Most common remote sensing systems operate in one or several of the visible, infrared and microwave portions of the electromagnetic spectrum (Fig. 1).

Radio waves : These are the longest wavelength (lowest frequency) radiations of the EMS. The wavelength of radio waves is more than 100 cm. These waves pass through Earth's atmosphere easily. Radio signals are used in radios, televisions, aircrafts, ship etc.

Micro-waves : Their wavelength ranges between 1 mm and 1m. RADAR (Radio Detection and Ranging) is the most common device used in Microwave Remote Sensing.

Infrared Radiations: When we feel hot it is because of infrared (IR) radiations. For common understanding we can call them as 'Heat'. The wavelength of IR is longer than visible light and shorter than microwaves approximately ranges between 1 micron to 100 microns. These are very useful radiations for remote sensing. Thermal Imaging Systems detect objects by recording their temperature (infrared-emissions).

Visible Radiations: As their name itself suggests these are the EMR (Electro Magnetic Radiations) which are visible to our eyes in different colours. These range between 700 nm to 400 nm. Most of the remote sensing systems and camera records images in this range.

Ultraviolet (UV) Radiations : These radiations have wavelength shorter than violet colour of visible light and longer than X-rays. UV radiations can be divided into near UV (400–200nm), far UV (200–10) and extreme UV 1–31 nm.

X-rays : These are very short length electromagnetic radiations wavelength in the approximate range from 0.01 to 10 nanometers. In EMS these fall between UV radiations and Gamma-rays.

Gamma-rays : Gamma-rays are the electromagnetic radiations with shortest wavelength in the range of the range of 10^-11 m to 10^-14 m. Because of very high energy these rays can cause serious damage to living cells.

Recognition of land surface features through remote sensing platform requires knowledge of surface reflectance which is also called the spectral reflectance, defined as the reflectance characteristics of earth surface features (land, soil water and vegetation) may be quantified by measuring the portion of incident energy that is reflected.

This is measured as a function of wavelength and is called spectral reflectance, rl = ER (l) /EI (l)

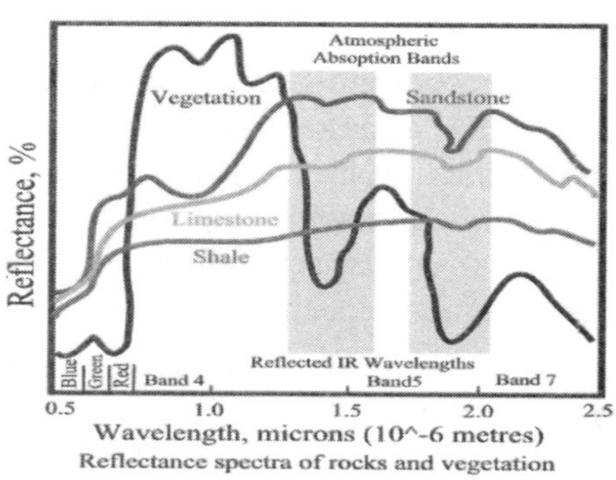

Fig. 1. Spectral curve of vegetation and rocks

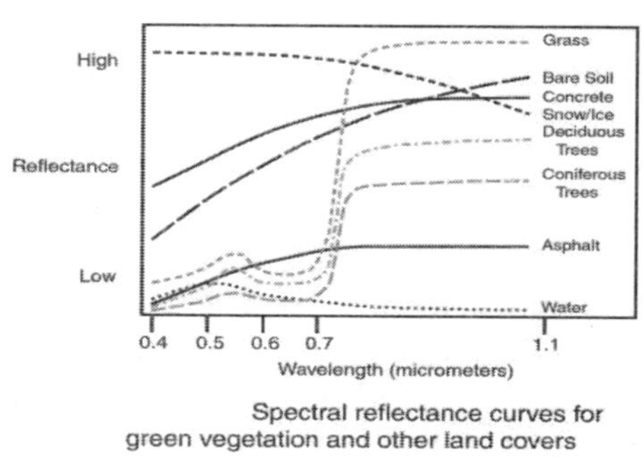

Fig. 2. Spectral curve of green vegetation and other surfaces

Where ER is reflected energy and EI is incident energy. A graph of the spectral reflectance of an object as a function of wavelength is termed as spectral reflectance curve (Fig. 1 & 2) : Source : http://instruct.uwo.ca/earth sci

Spectral Reflectance of Land surfaces

Spectral-Reflectance-of-Vegetation varies with wavelength. A compound in leaves called chlorophyll strongly absorbs radiation in the red and blue wavelengths but reflect green wavelength. The internal structure of healthy leaves act as diffuse reflector of near-infrared wavelengths.

Spectral-Reflectance-of-Water : Majority of the radiation incident upon water is not reflected but either is absorbed or transmitted. From the Fig. 2, one can see how the spectral curve for water is too low amongst all the other surfaces. Longer visible wavelengths and near-infrared radiations are absorbed more by water than the visible wavelengths. In the images, water surfaces look either blue or blue-green due to stronger reflectance at these shorter wavelengths and darker if viewed at red or near-infrared wavelengths.

Spectral-Reflectance-of-Soil: The majority of radiation on a surface is either reflected or absorbed and little is transmitted. The characteristics of soil that determine its reflectance properties are its moisture content, texture, structure iron-oxide content. The soil curve shows less peak and valley variations. The presence of moisture in soil decreases its reflectance.

Multi-spectral vs hyper spectral images: The multispectral satellite sensors contain less number of bands with broad spectral band width hence these are not capable of detecting fine details of the earth features being sensed. The broad band width cannot separate the objects having very little difference in their spectral reflectance. On the other hand, Hyper-spectral Imaging (HSI) instruments, possessing narrow large number of bands, make it possible to differentiate between the objects/features which may look similar in multispectral sensors. HSI can afford the detection of small alterations in the moisture content of the leaves, moisture status of soil, nutrient stress and other environmental stresses in the plants.

Satellite sensors for drought assessment

So far two coarse resolution satellite sensors (as mentioned below), are being used routinely to monitor vegetation and detect the impact of moisture stress on vegetation.

1. *Advanced Very High Resolution Radiometer (AVHRR).* These are mounted on board the National Oceanic and Atmospheric Administration's (NOAA'S) polar-orbiting satellites. The advantage are; collection of coarse-resolution imagery worldwide due to its twice-daily coverage and synoptic view.

2. *MODIS data* : In December 1999, NASA launched the Terra spacecraft, the flagship in the agency's Earth Observing System (EOS) program. The Moderate-resolution Imaging Spectroradiometer, or MODIS, is a sensor mounted on Terra that greatly improves scientists' ability to measure plant growth on a global scale. In comparison to AVHRR's almost-daily global cover, MODIS provides much higher spatial resolution (up to 250-meter resolution). In other words, MODIS will provide images over a given pixel of land just as often as AVHRR, but in much finer detail. Secondly, it has the advantages of greater number of wavelengths for measurement of land surface dynamics.

Remote sensing data availability under various platforms for drought assessment

In the following, a list of data products is provided in the Table 3 which may help in drought related research.

Table 3. Data products available under different platforms

Data products	Application Description	Spatial Resolution	Temporal resolution	Data availability
MYD09GQ Aqua	Surface Reflectance Bands 1–2	Tile	250m	Daily
MOD09GQ Terra	Surface Reflectance Bands 1–2	Tile	250m	Daily
MYD11C3 Aqua	Land Surface Temperature & Emissivity	CMG	5600m	Monthly
MYD11B1 Aqua	Land Surface Temperature & Emissivity	Tile	5600m	Daily
MYD11_L2 Aqua	Land Surface Temperature & Emissivity	Swath	1000m	5 min
MOD13Q1 Terra	Vegetation Indices	Tile	250m	16 day
MYD15A2 Aqua	Leaf Area Index - FPAR	Tile	1000m	8 day
MYD13C2 Aqua	Vegetation Indices	CMG	5600m	Monthly
MOD13C2 Terra	Vegetation Indices	CMG	5600m	Monthly
MYD13C1 Aqua	Vegetation Indices	CMG	5600m	16 day
MOD13C1 Terra	Vegetation Indices	CMG	5600m	16 day
MYD13A3 Aqua	Vegetation Indices	Tile	1000m	Monthly

MOD13A3 Terra	Vegetation Indices	Tile	1000m	Monthly
MYD13Q1 Aqua	Vegetation Indices	Tile	250m	16 day
MYD13A2 Aqua	Vegetation Indices	Tile	1000m	16 day
MOD13A2 Terra	Vegetation Indices	Tile	1000m	16 day

Monitoring vegetation through Vegetation indices

As has been mentioned earlier, remote sensing and geospatial technology are helpful in detecting and mapping spatial variations in the vegetation and the stress which are indicators for drought monitoring. In the following few of the indices have been listed. It is a fact that crop/vegetation reflects high in the near infrared due to its canopy geometry (Fig. 1). Within these bands (near infrared, red and middle infra-red bands), the reflectance curve is indicative of the health and conditions of crops.

- Difference Vegetation Index
- Soil Adjusted Vegetation Index
- Weighted Difference Vegetation Index
- Greenness Vegetation Index
- Normalized Difference Vegetation Index (NDVI)
- Enhanced Vegetation Index (EVI)

Out of these NDVI and EVI are the most frequently used indies, they are discussed as under

Normalized Difference Vegetation Index (NDVI): It derived from AVHRR data, has been extensively used for vegetation monitoring, crop yield assessment, and drought detection.

The NDVI is calculated as (NIR - RED)/(NIR + RED), where NIR is the reflectance radiated in the near-infrared waveband and RED is the reflectance radiated in the visible red waveband of the satellite radiometer. Higher NDVI indicates a greater level of photosynthetic activity (Tucker et al. 1991). It has been demonstrated that multi-temporal NDVI derived from AVHRR data is useful for monitoring vegetation dynamics on a regional and continental scale. NDVI products can be generated from the data of following satellite sensor systems; MODIS NDVI of 250 m and 1000 m, SPOT VGT NDVI of 1000 m. NOAA AVHRR NDVI of 1000 m, IRS WiFS NDVI of 188 m and AWiFS NDVI of 56 m are widely used for drought monitoring.

Principle : Crop canopy would vary during its stages of growing / decaying. While a peak would be expected during a peak vegetation cover, the fall in the curve would be expected during decay condition. The VI (Vegetation Index) peaks can therefore, be related to biomass of crops, biomass in the crop field etc.

A fall in the VI values would reflect a reversal, indicating crop stress. The reasons may be; inadequate moisture, arising out of prolonged rainfall deficiency /drought or may be due to infestation of disease. In general, comparison of VI profile of the reporting year and a previous normal agricultural year provides assessment of drought impact in the scale of previous agricultural scenario.

Enhanced Vegetation Index (EVI) : The MODIS Science Team is preparing a new data product–called the Enhanced Vegetation Index (EVI) that will improve upon the quality of the NDVI product. The EVI will take full advantage of MODIS' new, state-of-the-art measurement capabilities. While the EVI is calculated similarly to NDVI, it corrects for some distortions in the reflected light caused by the particles in the air as well as the ground cover below the vegetation.

Satellite Based Drought Assessment and Warning in India and role of NADAMS

In our country, NRSC (National Remote Sensing Centre) at Hyderabad, runs a project called National Agricultural Drought Assessment and Monitoring System (NADAMS) since 1990, which envisages use of remote sensing and meteorological data for drought assessment of 13 agricultural important states of India at three levels; state, district and sub district level. As per its report-2012, district-wise analysis in 13 states by the end of July indicated "Normal" agricultural situation in 132 districts. The agricultural situation is categorized as "Watch" in 148 districts and as "Alert" in 119 districts (NRSC, 2010). Drought assessment is carried out on fortnightly basis and reporting of information to the User departments is done on monthly basis. Resourcesat 1/ Resourcesat 2 AWiFS based sub-district level agricultural drought assessment is carried-out in 4 states namely Andhra Pradesh, Karnataka, Maharashtra and Haryana. AVHRR/MODIS based district level assessment is carried out in 9 states namely Bihar, Chhattisgarh, Gujarat, Madhya Pradesh, Orissa, Rajasthan, Uttar Pradesh, Jharkhand and Tamil Nadu. From the year 2012, the NADAMS project is being carried out at Mahalanobis National Crop Forecast Centre (MNCFC), Ministry of Agriculture (GOI), with the technical support of NRSC.

Approach: The methodology of NADAMS project essentially reflects the integration of satellite derived crop condition / surface wetness indices and their anomalies with ground collected rainfall and crop area progression to evolve decision rules on the prevalence, intensity and persistence of agricultural drought situation. Following multiple indices are used in agricultural drought assessment.

- Normalized Difference Wetness Index
- Normalized Difference Vegetation Index

- Soil Moisture Index (derived from soil water balance approach)
- IMD Rainfall data – rainfall deviation,
- Number of dry weeks

In the following table, details of procedure adopted by NADAMS for this warning are presented.

Table 4: NDVI based warning schedule

Month	Assessment	Implications
June, July, August	Normal	Agricultural situation is normal
	Watch	Progress of Agricultural situation is slow Ample scope for recovery No external intervention needed
	Alert	Very slow progress of agricultural situation Need for intervention Develop and implement contingency plans to minimize loss
September, October	Mild drought	Crops have suffered stress slightly
	Moderate drought	Considerable loss in production Take measures to alleviate suffering
	Severe	High risk. Significant reduction in crop yield Management measures to provide relief

NDVI DISTRIBUTION OVER JAISALMER DISTRICT IN WESTERN RAJASTHAN DURING DROUGHT AND NORMAL YEARS

NDVI or Normal Difference Vegetation Index are the commonly used methods adopted to analyze remote sensing measurements and assess whether the target being observed contains live green vegetation or not. A NDVI based analysis was carried out to assess the state of vegetation cover in Jaisalmer district of Rajasthan. The study used NASA MODIS CMG Monthly NDVI (Global) data from the Terra sensor projected on a 0.05 degree (5.5 km resolution). IRS - AWiFs data were used as secondary data for checking vegetated area. The NDVI data were downloaded for the years from 2000 to 2012 and for monsoon period (July, August and September). For generation of yearly average scenes, raster calculator was used to calculate total values for all three seasons averaged by 3 for each year. The maximum and minimum range of NDVI values fell within a range of 0 to 0.65. These values were further classified into three categories (0.65/3) as 0-0.2 (Low), 0.2-0.4 (Moderate) and 0.4-0.7 (High) using ArcGIS spatial analyst. Figure 3 presents average NDVI maps for three situations (more than normal, normal and deficit conditions of rainfall). The year 2000 represented more than normal year, 2003 (succeeding the 2002 severe drought year) and 2009 (normal year). These images present an over view of spatial changes in greenness in the region.

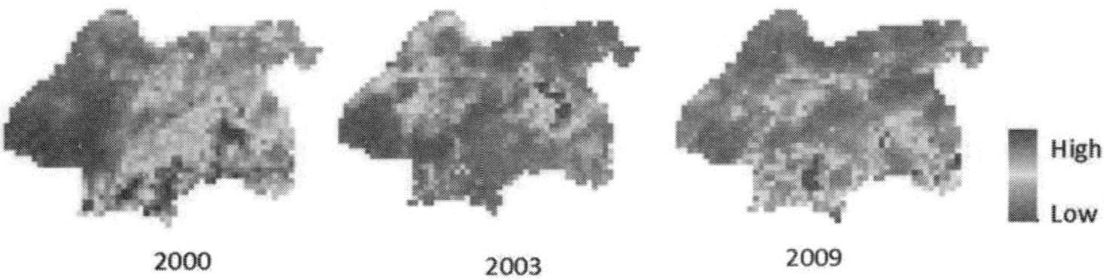

Fig. 3. NDVI derived from MODIS TERRA showing different levels of Greenness over Jaisalmer district from 2000 to 2009.

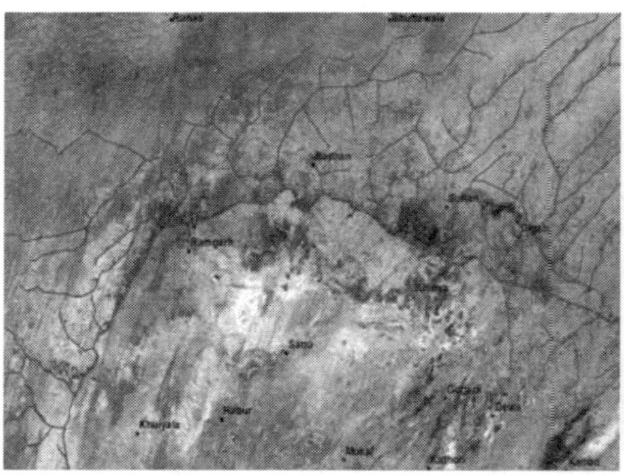

Fig. 4. Irrigated IGNP canal catchment, Jaisalmer on IRS LISS III FCC, 2014, Source : Author

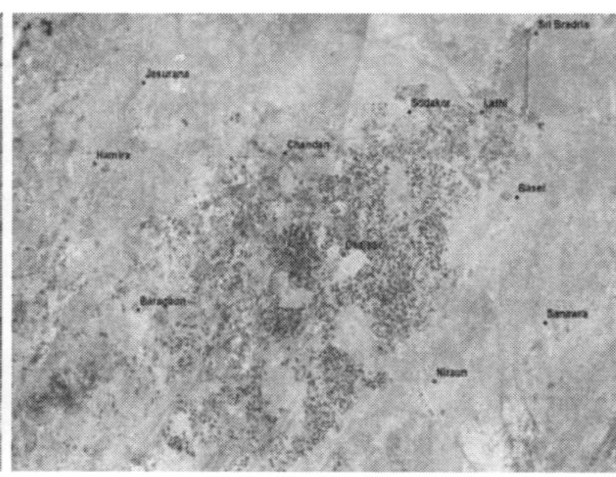

Fig. 5. Tube well irrigated areas, Jaisalmer, on IRS LISS III FCC, 2014, Source : Author

Assessment: From the above yearly NDVI images, it is indicative that the central and western part of the district has very poor vegetation cover. Studies based on survey of natural resources (Chatterji and Kar, 1992) and also based on geomorphology maps (Moharana et al 2013), indicate dominance of aeolian landforms which includes > 75% area under sand dunes in the south and south western part. Studies on land use/land cover changes between 1980 and 2006 by scientists at CAZRI (CAZRI, 2008) indicate substantial increase of rainfed cropland by 332%, permanent pasture by 105% while decreasing area under watselands by 26.9%. Such data indicate how situations of agriculture have improved in several parts of an arid district in Rajasthan. The increase in irrigated croplands are now clearly visible in the FCCs (False Color Composites) of satellite images (as can be inferred from figures 4 & 5, from IRS LISS III satellite images). The rainfed lands are now intensively utilized as irrigated croplands in the lower catchment of IGNP canal between Mohangarh and

Ramgarh (Fig. 4). In this image such areas are depicted by red tone. Similarly, higher NDVI is also noticed in scattered patches in the tube well irrigated regions near Devikot and Sankra (Fig. 5) in the southern part of the district. Such irrigated croplands can be easily identified on the basis of the pattern of croplands. The vegetation cover is in the form of few trees (*Prosopis cineraria*, (Khejri) *Prosopis juliflora* (Babul), shrubs (*Calligonum polygonoides* (Fog), *Leptadenia pyrotechnica* (Khimp), *Capparis decidua* (Ker), *Ziziphus nummularia* (Bordi) and *Ziziphus mauritiana* (Ber) and good grass covers like *Lasiurus sindicus*, *Panicum antidotale*, *Panicum turgidum* (Narain et al., 2005).

CONCLUSION

Drought is a common phenomenon in arid western part of Rajasthan where incidences of drought put considerable impact considering the demand of food and fodder for its 19.8 million human and a matching livestock population. Remote sensing based information derived from both coarse and moderate resolution satellite sensors and the drought indices products are useful to routinely monitor vegetation and crops as indicator of drought severity, particularly for arid region like western part of Rajasthan. The region, very much depend on the occurrence of seasonal rainfall, and hence more vulnerable to droughts. Besides conventional approaches to drought assessment and warning, satellite based information provides a quick input to understand and predict the state of climatic abnormalities and so as various types of droughts. The study includes a case study of remote sensing based approach to see the variability in the vegetation density cover in Jaisalmer district of Rajasthan. The variability in the NDVI and its further classification showed vulnerability of this region to rainfall deficiency, vegetation poor growth and hence to drought situations in a temporal variation from 2000 to 2012.

REFERENCES

Albert J Peters, Elizabeth A Walter-Shea, Lel JI, Andres Vliia, Michael Hayes, and Mark D Svoboda 2002 Drought Monitoring with NDVI-Based Standardized Vegetation Index, *Photogrammetric Engineering & Remote Sensing.* **68(1)** 71-75.

CAZRI 2005 Drought in western Rajasthan, Impact, coping mechanism and management strategies (eds) Pratap Narain and Amal Kar, NAIP, CAZRI. 104 page.

CAZRI 2008. Annual report, Central Arid Zone Research Institute, Indian Council of Agricultural Research. Jodhpur. 166 pages

Chatterji PC and Amal Kar 1992 Integrated Natural and Human resources Appraisal of Jaisalmer district. Central Arid Zone Research Institute, Jodhpur 156 pages.

Dai A 2011 Drought under global warming. Wiley Interdisciplinary Review*: Climate change* **2(1)** 45-65

Moharana PC, Gaur MK, Choudhary C, Chauhan JS and Rajpurohit RS 2013 A system of geomorphological mapping for western Rajasthan for its relevance in Agricultural Land use. *Annals of Arid Zone* **52(3&4)** 163-180

Narain P, Manjit Singh, MS Khan and Suresh Kumar 2005 *Shrubs of Indian Arid Zone* (eds) Arid Agroecosystem Directorate, NAIP, CAZRI. 176 pages.

NRSC 2010. *Remote sensing applications* (eds) PS Roy, RS Dwivedi and D Vijayan, National Remote Sensing Centre, Hyderabad. 377page

Saha DK, Kar Amal and Roy MM 2012 Indicators of drought vulnerability for assessing coping mechanism in arid western Rajasthan. *Annals of arid zone* **51(1)** 1-9

Subbaiah AR, 2004. *State of the Indian Farmer- A Millennium study*, Natural Disaster Management, 21, Department of Agriculture and Cooperation, Ministry of Agriculture, Government of India, New Delhi, 226 page.

Tucker CJ, Dregne HE and Newcombe WW 1991 Expansion and contraction of the Sahara Desert from 1980–1990 *Science* **253** 299–301

6

Drought Vulnerability and Adaptation

K.D. Sharma

INTRODUCTION

The Indian Ocean provides moisture for the vast southwest summer monsoon current over the Indian landmass, whereas the Bay of Bengal feeds the northwest winter monsoon restricted to only the southern India. The Northwest India is extremely dry receiving annual rainfall less than 500 mm. These desertic and semi-arid conditions are in marked contrast to the tropical monsoon and very wet rainfall regimes (1000 - 2000 mm) in the coastal region. The whole India experiences water deficit in non-rainy periods. Year to year variation in agricultural productivity throughout the region is related to variations in rainfall. Traditional agricultural practices continue to dominate most of the region. Disastrous droughts and floods have historically caused crop failures, which has led to strained economies and abnormal food shortages; at times famine like situations leading to hunger and malnutrition for the vulnerable communities. Continuing increase in human and livestock population further add to the problem. In lowland areas (<300 m), rice-wheat (irrigated) is the main production system. In the intermediate elevation zone (300 – 1500 m), upland rice, maize, millet, sorghum, cotton, pulses, oilseeds, rubber, oil palm, coffee, sugarcane, fruits and vegetables are grown. The discussion in this chapter is centred on the issues of drought mitigation in this intermediate elevation zone.

As per current theory, the drought is identified as a phenomenon of temporary negative deviation in environmental moisture status. Because of the inherent variable character of monsoon rainfall there are often occasions when the actual rainfall is much below the 'normal' expected on a long-term basis. This results in drought, which is manifested in terms of crop failure and depletion

---

Former Member and Technical Expert (Water Management),
National Rainfed Area Authority, Planning Commission, New Delhi 110012
E-mail: drkdsharma@gmail.com

in surface and groundwater resources/water bodies. The consequence is large-scale human migration, loss of livestock and human lives due to scarcity of water, food and fodder (Oliver and Fairbridge 1987).

CAUSES OF DROUGHT

The root cause for deficit rainfall that results in drought is the widespread and persistent atmospheric subsidence arising from the general circulation of the atmosphere. Recent studies on interactions between global circulations and drought showed that the *El Nino* phase of the Southern Oscillations (ENSO) contributes substantially to summer droughts adversely affecting the food grain production. The frequency of drought is also influenced by climatic change as a result of increased concentration of the atmospheric CO_2, methane and nitrous oxide. The indiscriminate use of gases like chlorofluoro carbon (CFC) and use of coke in thermal power plants have altered the radiation balance of the atmosphere resulting in increased temperatures due to Green House effect. The Inter-governmental Panel on Climate Change (IPCC) of the World Meteorological Organisation projected an increase in atmospheric temperature of the order of 0.1 to 0.3°C by 2010 and 0.4 to 2.0°C by 2020 in South Asia that may result in a decrease in cereal production by 5 to 15% in the region (Houghton et al 1990). Also, the extensive deforestation in vogue since centuries has altered the hydrological cycle thus enhancing the incidence of droughts in the country.

FREQUENCY OF DROUGHTS

A study of moderate and severe droughts that occurred in India indicates that, except for very small pockets in the North eastern India and Kerala, there were no areas, which have not been affected by drought at one time or the other. While the entire country could thus be considered as drought prone, there are certain areas, which are chronically subject to such condition and merit the appellation 'drought prone' (Table 1). Technical Committee on Drought Prone areas Programme (DPAP) and Desert Development Programme (DDP) identified about 120 million ha of the country's area, covering 185 districts (1173 development blocks) in 13 states as drought prone (Anonymous 1994). Based on the historical records, Jaiswal and Kolte (1981) reported 120 drought/famine like incidences in one or other part of the country between 1291 and 1979. During the 20th century alone, droughts of varied intensities occurred during 28 years in India (Venkateswarlu 1997).

IMPACT OF DROUGHTS

The usual impact of agricultural drought is in terms of loss of crops, malnutrition of human being and cattle, land degradation, loss of other economic activities, spread of diseases, and migration of people and livestock (Kulshrestha 1997). Droughts result in crop losses of different magnitude, depending on their geographic incidence, intensity and duration. The droughts not only adversely affect the food security at the farm level but the national economy and overall food security as well. Predicted losses to agriculture in India were 50% during the drought of 1957-58. The drought of 2002 resulted in 25% and 16% reduction in rice and oilseed production, respectively. Ramakrishna and Rao (1991) observed that during the 1987 drought in India, the productivity of pearl millet dropped by 78, 74 and 43% in rainfall zones of <300, 300-400 and > 400 mm, respectively. Victor et al (1991) reported similar reduction in the productivity of groundnut and millet in Aṇdhra Pradesh during the drought. For eastern India, the loss in production of food grain due to drought averaged over 1970-96 has been estimated to be $ 400 million year^{-1}, which is equivalent to 8% of the value of food grain production in the region (Pandey et al 2000). The effect of drought was more pronounced on fodder availability as compared to that of food grains.

Table 1. Probability of occurrence of drought in different meteorological subdivisions of India

Meteorological subdivision	Frequency of deficit rainfall (75 % of normal or less)
Assam	Once in 15 years
West Bengal, Madhya Pradesh, Konkan, Bihar and Orissa	Once in 5 years
South interior Karnataka, Eastern Uttar Pradesh and Vidarbha	Once in 4 years
Gujarat, East Rajasthan, Western Uttar Pradesh	Once in 3 years
Tamil Nadu, Jammu and Kashmir and Telangana	Once in 2.5 years
West Rajasthan	Once in 2 years

The duration of availability of water in surface water bodies reduces significantly during drought year (Narain et al 2000). They get dried up quickly even before the onset of summer. The groundwater table declines and the shallow wells dry up. Sometimes, the concentrations of toxic elements such as arsenic, fluoride and nitrate are increased. In deep wells, concentration of salts increases due to lowering of water table. The poor people suffer the most since they own shallow wells and cannot afford to deepen them.

Droughts affect the livestock in several ways. Reduced productivity and mortality are the direct effects. Driven by enhanced livestock pressure due to depletion of forage resources during drought, overgrazing and indiscriminate cutting of vegetation takes place leading to land degradation. This is followed first by distress sale of cattle and even small ruminants. Migration is the next step extending the problems of uncontrolled and overgrazing, thus degrading land in other areas. There have been instances that large-scale mortality of livestock and mismanagement in disposal of their carcass caused epidemic situations and environmental hazards. Decrease in size of herd (up to 52%) was reported due to frequent occurrences of droughts (Anonymous 1994).

Recurring droughts bring about changes in the nature and extent of the socio-economic values and attitude of the people (Purohit 1993). Rise in prices, contraction of charity, diminution of credit and consequently enhancement of the rate of interest on loans, reduced grain trade, increase in petty crimes, unusual migration of the people along with their herds, etc. were some of the most common outcomes. The agricultural sector no longer could provide basic employment to rural workers thus resulting in migration/underemployment, and creation of slums in the cities. Rural women were overburdened with additional work to earn through relief works. The infant mortality was high succumbing to host of diseases mostly due to drinking poor quality water. Taking infants to work/migration also led to increased child mortality. The mother and child had high morbidity due to low intake of calories, much below the energy spent for hard labour in relief works.

Apart from the natural causes (lack of rainfall) and the antecedent conditions (climatic characteristics, soil conditions, presence of groundwater), the interaction of the inhabitants (human and livestock) has profound impact on the growth of drought. It is not that only the farmers are adversely affected by drought but the entire society suffers due to rise in prices and additional taxation, to meet the national needs for import of food. However, the farmer is the greatest sufferer for it is this sector of society, which bears the brunt in terms of loss in production. Hence, there is a need for action to protect farmers from devastation caused by the droughts, primarily through national and also regional and international interventions.

It must however be clearly understood that the impact of drought is not just a temporary adversity or a calamity. The relief measures, imports, foreign assistance and taxation on middle class society to meet the food needs are merely the national attempts to address to the secondary impacts of drought. The basic adverse impact of drought is obviously the continuing land degradation or desertification in vulnerable areas, which if continued, can become irreversible in not too distant future. Obviously, there are many direct factors too, leading to land degradation. Yet since drought has a direct impact on agricultural production, any effort aimed at control of land degradation and protection of natural resources, would substantially mellow down the adverse impact of drought on agricultural productivity.

The scenario of drought discussed above thus, reflects the drought as a growing threat to the very fabric of the society and the environment. Therefore wide ranging policy, infrastructural and technological issues need to be looked into more seriously than ever in national, regional and global perspectives involving Governmental, Intergovernmental agencies and mechanisms. The national agricultural research systems (NARS), International Institutions, NGOs and Private Sector need to work together to realize the goal of a sustainable agricultural enterprise based on the most efficient use of the available natural resources i.e. land, water and vegetation in an integrated manner and in a farming systems perspective.

DROUGHT MITIGATION STRATEGY

Drought is indeed a many faceted natural disaster that leads to serious socio-economic impacts, which have long-term implications. It is in this context that development of appropriate drought management strategy is of great importance, for individual countries as well as through international initiatives. Drought management is currently addressed by the following mechanisms and sectors:

Governmental

- Policy issues, national, regional, and district level
- Rural development infrastructure
- Input supply, marketing and farm advisory services

Non-Governmental

- NGOs
- Rural institutions, local self-governments
- Private sector
- Philanthropic organizations
- Community codes (tribes, herders)
- International aid agencies

Research and Development Institutions

- Weather forecasts, early warning, vulnerability and preparedness measures
- Best practices for rainwater and soil management through linking on-station and on-farm research
- Contingency crop planning/mid-season corrections
- Alternate land use systems

Research and development initiatives especially in areas like understanding the monsoon behaviour, agro meteorology, arid/dryland farming systems and hydrology have since been contributed substantially to the knowledge base on drought management. These advancements have contributed in the development of useful technological options and also infused dynamism in agricultural production strategies and development of appropriate farming systems. These efforts thus enable the farmers to tide over difficulties created by the drought situations. However, the technology adoption by small holders who constitute the bulk of stakeholders in drought prone areas has not been to the desired extent. This shortfall is not only due to inherent difficulties in developing appropriate technologies for extremely complex rainfed–drought prone environment, but also due to the poor socio-economic conditions of the farmers and lack of adequate development infrastructure. Of late, the drought management approach has shifted significantly in the region from crisis response to risk management through early warning systems, advance planning for emergency response and better preparedness. This paradigm shift in R&D strategy has since made some impact, but the sustainability in rainfed areas is yet to be achieved. The quality of life of small holders in rainfed areas therefore, is yet to attain the minimum level of the desirable standards of living to reduce the gap with high productivity zones i.e. the irrigated areas. This is a national priority.

Having gone through the unnerving experience of suffering many widespread droughts due to high variability of rainfall in time and space, a fairly organized drought management system has evolved in last 150 years in India. The problems caused by drought were faced with fair amount of competence and dedication. However, the approach was essentially that of starting *ad hoc* relief works, as per famine code, to provide employment to the distressed population rather than having the long-term perspective for improving the conditions in the region and to reduce the probability of drought. This approach did not prove effective in mitigating the drought conditions. It was only since 1970s, with the launch of Drought Prone Area Programme (DPAP), that a long-term view was taken for evolving technological and organizational innovations, which were accepted as the strategy for integrated development of drought prone areas. This policy improvement has paid rich dividends while dealing with the infamous droughts of 1987 and 1999-2000. At present, the main objectives of drought management strategy are (Anonymous 1994):

- Distribution of essential commodities such as water, fodder and food at subsidized rates, sometimes under Food for Work Programmes.
- Optimum utilization of resources in the affected areas with emphasis on primary resources viz. soil, water, vegetation, livestock, man-power, etc. Relief works undertaken for providing employment to drought stricken population are mainly for drought proofing.
- Improvement in terms of living conditions of the rural poor who suffer most due to scarcity and drought in particular, and the community in general by creating direct and indirect wage employment and taking up short gestation programmes of development.

In India, major research efforts on improving the productivity of rainfed areas with focus on reducing the adverse effects of drought have been underway for the last 2-3 decades. Package of practices involving appropriate crops, improved varieties, tillage and seeding practices, soil and rainwater conservation /harvesting, use of fertilizers, weed control, alternate land use systems and plant protection have been developed through intensive and location specific research efforts. However, except for improved crop varieties, other components of the technology – particularly those aimed at conservation of natural resources (soil and water) have not been adopted by the farmers, more so the small holders who dominate the agriculture sector in rainfed areas. The integrated watershed management programme which forms the central strategy for the national programme of development of drought prone areas in India, initiated in mid-seventies has not made desired impact in terms of replicability of pilot project despite the huge investment of public funds during the last two decades. Among other reasons, the following factors have been responsible for this setback:

- Technology bias on biophysical issues – lack of appreciation of farmers conditions, their priorities and resources
- Top down extension strategy
- Lack of on-farm research for participatory technology development (PTD) in rainwater harvesting and watershed management
- Greater reliance on crops – lack of appreciation of farming system perspective and production system diversification
- Inadequate rural infrastructure

As part of its Medium-term Plan (2002-2007), FAO is improving inter-departmental coordination through defining disaster prevention, mitigation and preparedness, and post-emergency relief and rehabilitation as one of the five Priority Areas for Inter-disciplinary Action (PAIA Figure 1). FAO approach to disaster management (FAO 2002) including tropical cyclones, floods, droughts and earthquakes is in line with the United Nations international strategy for disaster reduction. FAO further described interventions in emergencies in terms of a sequence of events referred to as a disaster cycle, with distinct phases, each requiring different action. The eight type of action or phases of the emergency sequence are: (1) prevention, (2) preparedness, (3) early warning, (4) impact and needs assessment immediately following a disaster, (5) relief, when immediate humanitarian assistance is required, (6) rehabilitation, when the first attempts to rebuild the rural livelihood system take place, (7) reconstruction, when the destroyed infrastructure is replaced and investment can take place, and (8) sustainable recovery, when conditions permit to return to a development process. The key areas for strategic intervention under each action/phase are depicted in Table 2.

Table 2. Proposed measures under operational strategies for disaster management in the Asia-Pacific Region

Prevention	**Preparedness**	**Early Warning**	**Impact and need assessment**	**Relief**	**Rehabilitation, reconstruction and sustainable recovery**
Introducing Food Insecurity and Vulnerability Information Mapping Systems (FIVIMS)	Promote appropriate database for preparedness and planning	FIVIMS	Establishing methods for community and government impact and needs assessment	Improving collaboration of all players at local, national, regional and global levels	Integrating all actions with the development process
Strengthening guidelines for agricultural vulnerability reduction	Support disaster preparedness planning in farming systems	Establishing/ strengthening national EWS	Improving comprehensiveness, timeliness and accuracy	Increasing cost-effectiveness of distribution mechanisms	Building physical and institutional infrastructure
Encouraging hazard resistant and protective structures	Strengthen Asian, national and local level FIVIMS	Establishing/ strengthening agro-meteorology units	Promoting community and government participation	Mobilizing resources (foods, inputs, funds and support)	Raising R & D for resilient farming systems
Promoting land and water use planning and zoning	Create mechanism for disaster management	Upgrading crop assessment and forecasting methods	Linking national, FAO/ WEP and donor assessments	Instituting nutrition surveillance	Strengthening agricultural micro-credit systems
Developing community decision-making and action	Support nutritional information and health care	Linking national, regional and global EWS		Building systems for monitoring and evaluation	Improving access to land and other means of production
Building community self-reliance	Develop mechanism for loss and damage compensation	Improving two-way farmer-EWS information flow			Raising farm and off-farm incomes of the poor
* Improving disaster proof farming systems including: - Crop diversification - Integrated farming systems - Contingency crop planning - Disaster mitigation practices - Hazard resistant practices - Pilot projects	* Stock emergency food, feed, fodder and agricultural inputs * Encourage agriculture-related employment projects * Work towards effective involvement of local communities				

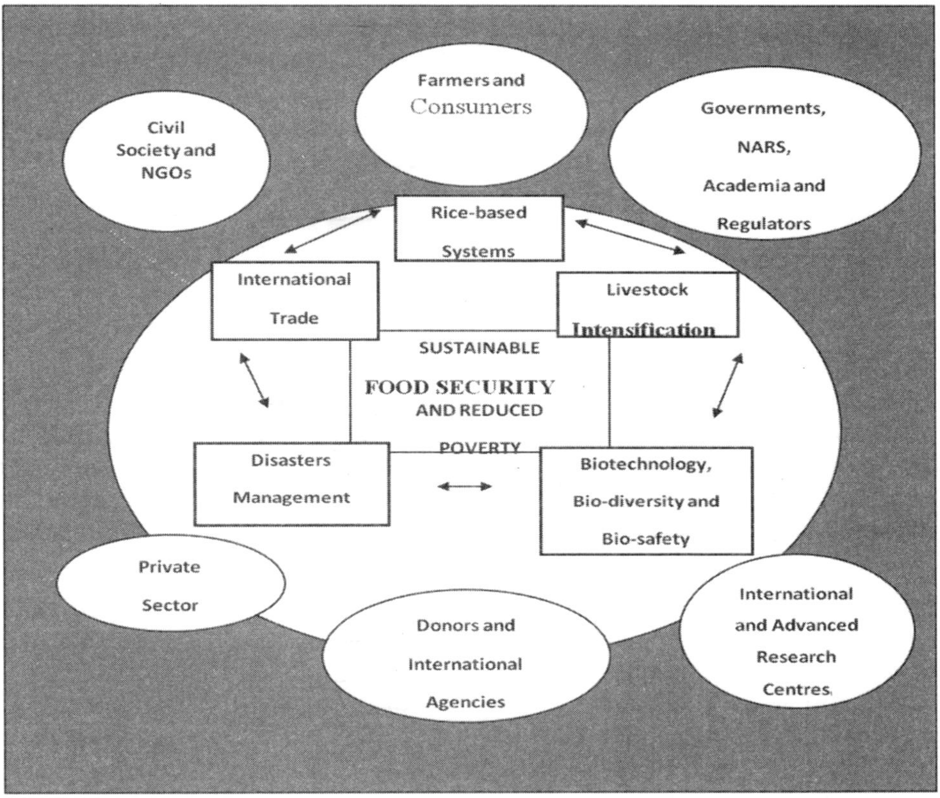

Fig. 1. Stakeholders, and priority inter-disciplinary actions, to strengthen food security, lessen poverty, and sustain environments

FARMING SYSTEMS PERSPECTIVE FOR DROUGHT MITIGATION

Extension and development programmes have failed to inspire farmers to adopt technologies in drought prone areas. Besides lacking a farmer participatory approach as stated earlier, these failures partly stem from the inherent difficulty of the task. Of late, it has been increasingly recognized how difficult it is to develop sustainable technology for heterogeneous agro-ecological and socio-economic conditions of small holders in drought prone areas, unlike that for irrigated areas. The problems are complex, characterized by a host of environmental and socio-economic issues. Addressing only a component of the farming system, e.g., crop variety, fertilizer use, or even crop husbandry *per se,* does not generally result in dramatic increase in productivity as witnessed in irrigated areas. Mixed farming, consisting of crop production and animal husbandry for risk aversion, has been the mainstay of subsistence for such farmers. The farming systems perspective, therefore, can only be the proper management strategy for such regions.

Planning on-farm farming system research requires an understanding of what farmers are already doing. Four sets of information are needed to initiate farming system research in real field situations (Singh 2001):

i. Socioeconomic conditions of people, their perceptions, priorities, requirements, and indigenous technical knowledge
ii. Natural resource conditions
iii. Technology (research information) developed at the research stations
iv. Infrastructure (market and input issues).

These data are collected and then carefully analyzed to balance resource conditions and farmers requirements for selecting technologies to apply and test as treatments. To begin with, more emphasis should be placed in refining indigenous technical knowledge. Strict prescriptions of component technologies developed at the research stations should be avoided. Needed adjustments can be incorporated to offer baskets of options that suit the client's requirements and priorities.

Sustainability of rainfed agriculture is anticipated to be one of the most crucial problems of the next two decades. In this context sustainable agriculture refers to a farming system, which on a continuous basis enhances the productivity and economic returns, protects the environment, conserves the natural resources, and finally leads to improved quality of life of people. The burgeoning human and livestock population regime in the region demands a continuous increase in productivity to meet the increasing requirements. Economic growth associated with development further adds to it. The fragile resource base in these regions is, therefore, required to be balanced with production activities in such a way that it continuously responds to the increased requirements without compromising its quality.

In irrigated areas the synergy among components of the recommended production technology (viz. high yielding variety, water management, fertilizer use and plant protection) resulted in a quantum jump in productivity. Farmers got impressed and readily accepted the technology, which led to the green revolution. However, in rainfed region, such synergy could at best operate at a very low level. Water, the most important input is uncertain. Moreover, farmers are poor and cannot cope with the risk factor. Inputs such as fertilizer are thus seldom used. In this back drops, getting the farmers interested in improved technology as willing participants becomes a challenge. Farmers will not accept any technology until productivity substantially increases. For this reason farmers in rainfed areas have not accepted technologies (in isolation) despite vigorous extension efforts during the past two decades. Crops alone cannot meet the formidable task of sustainable dryland farming. We therefore, have to look into the whole production/farming system for synergy among its components, e.g., arable cropping, livestock management, alternative land use systems, and management of village commons/degraded lands. With such a synergistic integration the farming systems approach can meet the objective of productivity increase to the desired extent in the rainfed areas.

STRATEGY TO CATALYZE FARMERS' PARTICIPATION

Technology development for rainfed areas in the past has by and large ignored the farmer. It is known that peoples' participation is hard to come by unless some tangible benefits come through in the immediate future.

The following methodology may hitherto be appropriate to obtain interactive participation of farmers in drought mitigation efforts:

- Participatory resource appraisal (PRA)
- Demonstration of land degradation processes and water and soil losses with tools such as portable rainfall simulator
- Focus group interactions to clearly obtain farmers' perceptions on: (i) drought prone issues; (ii) watershed and its relevance; and (iii) sustainability of rainfed farming
- Identification/listing of indigenous technical knowledge aimed at its utilization for on-farm assessment and refinement.

The success in interactive participation if focused in a mission-mode may eventually lead to voluntary participation or self-mobilization, which must be the national strategy for technology transfer in rainfed areas in the next decade. This methodology should invariably be adopted for planning on-farm drought prone research in a farming systems perspective.

- Manipulative participation
- Passive participation
- Participation for material incentive
- Functional participation
- Interactive participation
- Self-mobilization

RESEARCH AND DEVELOPMENT STRATEGY

Strategic planning for on-farm research, and Participatory Technology Development for developing best practices and farming systems is crucial and central to attain the goal of sustainable management of drought prone areas. In this context it would be pertinent to divide the on-farm program into short-term (5 years), medium-term (10 years) and long-term (20 years) technological interventions. Further, it is crucial to integrate these sets of interventions wherever possible on a watershed scale, otherwise on unit area basis, and initiate in a phased manner. This will result in not only ensuring peoples' participation right since the beginning of the on-farm research but also in sustaining and extending the approach to other areas, due to its demonstration value.

Short-term measures

Technologies to address the problems on short-term basis aim at immediate benefit to farmers. An ideal approach would be to start from refinements in indigenous technical knowledge based practices. These interventions would not need much monetary inputs yet may lead to tangible benefits from improved crop productivity, thus resulting in indigenous capacity building. This would build the confidence of the farmer in the programme and motivate them to adopt medium and long-term measures in due course, which will obviously need more of his contribution both in terms of labour and cash input. These consist of `doable' technologies, which the farmers can adopt at their farm level (irrespective of farm size) and of course, as far as possible on watershed basis with a little or no help from external agencies. Drought tolerant varieties, inter-bund treatments like key-line and appropriate tillage, summer tillage, ploughing/sowing across the slope, green-loppings-land cover-cum-manure treatment, vegetative barriers, ridge-furrow configuration for planting, opening of conservation furrows, etc. are some of the examples for short-term measures. These technologies aim at creating an impact within a short span of 2-5 years. Nevertheless, such technologies act as "starters" or "stepping stones" for the longer-term task of managing water stress by developing sustainable water resources. These interventions aim at ensuring utilisation of the *in situ* conserved moisture integrated with improved crop management for enhancing the crop yield under erratic and uncertain water supply. In summation, the core objective of the short-term measures is capacity building of farmers through the adoption of low input, doable technologies that triggers their willing participation in PTD.

Medium-term measures

Once the farmers begin to realize the benefits of adopting technologies (through short-term measures), they can better understand the goals for control of drought and respond to participate in the adoption of medium and long-term measures. The medium-term measures address the problems in a time frame of 5-10 years. Integrated watershed management involving PTD for development of farming systems and best practices is central to the medium-term measures. Regularization of runoff and storage in medium size reservoirs, renovation of existing tanks, adoption of alternate land uses systems and improvement of common pool resources are the examples of medium-term interventions. Construction of such structures needs major initiative from the farmer with support from the government. Best practices are those which are appropriate both from biophysical and socio-economic angles.

Long-term measures

These measures have a time frame of 10-20 years. Besides generation of large-scale surface and groundwater resources, development of wastelands which have gone out of cultivation should receive special attention in such interventions to reduce pressure on existing arable land resources for ease in drought mitigation. The emphasis on rehabilitation of wastelands should be to develop alternate land use systems with silvi-pasture integrated to livestock production as the main production enterprise. Extremely degraded and rocky terrains can be utilized as catchments for generation of surface water resource for utilization in adjoining areas. Such long-term measures thus can restore the land back to the production chain.

Long-term measures also include structures, which are erected to regulate overland flow and reduce peak flow. These structures, aim at improvement of relief, physiography and drainage features of watersheds on macro scale, say 2000-5000 ha.

CONCLUSION

Drought-prone areas in rainfed region are subjected to the ever-increasing pressure of population in India. Growing demand for food and fodder has been forcing the marginal lands to come under plough. Livestock suffers the most during drought for want of fodder and water. Climate change leading to increasing global temperature might lead to increased drought incidence in time and space. Human miseries are thus bound to increase if appropriate interventions to progressively mellow down this adversity of disastrous proportions, are not immediately initiated in order to effectively manage it over a time frame. In this backdrop, strategy to mitigate drought that may be adopted by the country can be summarized as following set of recommendations:

- Agro-eco-region/sub-region based problem prioritization to allocate resources and identify therein areas of researchable issues
- Watershed based rainwater and soil management, integrated nutrient management and integrated pest management for efficient utilization of natural resources and farm available inputs
- Simulation modelling using GIS, weather forecasts for early warning system and decision support systems
- On-farm research in a farming systems perspective for technology development, assessment and refinement in a participatory mode
- Integration of indigenous knowledge with scientific knowledge systems
- Agro-eco region (micro level) based developmental planning for short, medium and long-term interventions for rainfed drought prone agriculture

- Agro-biodiversity conservation
- R&D consortia development for all drought prone agro-eco sub-regions

The problems of drought-prone regions vary in magnitude, temporally and spatially. A multitude of organizations such as public/private sectors and NGOs are engaged in addressing drought related issues. These efforts are often sporadic and isolated thus do not lead to the desired level of success. Judicious use of scarce national resources is thus adversely affected. Therefore, there is a need for establishment of consortia for convergence of such scattered efforts.

From technological considerations, integrated watershed management with focus on land capability based production system diversification (agro-forestry, horticulture, etc.) and value addition is the best bet for sustainable management of resource poor drought prone areas. Agro-forestry not only reduces risk of crop failure but also helps in conserving and enhancing natural resources and protecting the all important environment resource for sustenance of life and its quality for both, the present and future generations. Further, rural agro-industry needs to be vigorously pursued not only for value addition to the produce, but also for creating rural employment in order to keep the required work force in place to practice agriculture not merely as livelihood but as economic enterprise for national food and nutritional security.

REFERENCES

Anonymous 1994 Report of the Technical Committee on Drought Prone Areas Programme and Desert Development. Ministry of Rural Development, New Delhi

FAO 2002 *Rural Asia-Pacific: Interdisciplinary Strategies to Combat Hunger and Poverty: 2000-2015*. Food and Agricultural Organisation of the United Nations, Regional Office for Asia and the Pacific, Bangkok, 92 Pages

Houghton JT, Jenkins G I and Ephraums JJ 1990 Climate change. *The IPCC Scientific Assessment WMO/UNEP*. Cambridge, UK

Jaiswal NK and Kolte NV 1981 *Development of Drought Prone Areas*. National Institute of Rural Development, Hyderabad

Kulshrestha SM 1997 *Drought Management in India and Potential Contribution of Climate Prediction*. Joint COLA/CARE Technical Report No. 1, Institute of Global Environment and Society, Calverton, USA

Narain P, Sharma KD, Rao AS, Singh DV, Mathur BK and Ahuja UR 2000 Strategy to Combat Drought and Famine in the Indian Arid Zone, Central Arid Zone Research Institute, Jodhpur

Oliver JE and Fairbridge RW 1987 The Encyclopedia of Climatology, Van Nostrand Reinhold Co., New York

Pandey S, Behura D, Villano R and Naik D 2000 Economic cost of drought and farmer's coping mechanism: a study of rainfed rice systems in eastern India. *Discussion Paper Series No. **39***, International Rice Research Institute, Manila

Purohit ML 1993 Socio-economic aspects of drought and desertification. In: *Desertification in Thar, Sahara and Sahel Regions,* Scientific Publishers, Jodhpur 205-218

Ramakrishna YS and Rao AS 1991 Incidence and severity of droughts in the Indian arid zone and their impact on productivity from agricultural and pasture lands. *Indo-Soviet ILTP Meeting on Ecology of Arid Zones and Control of Desertification,* Central Arid Zone Research Institute, Jodhpur.

Singh HP 2001 Developing farming systems and best practices for drought prone areas. In: *FAO Asia-Pacific Conference on Early Warning, Prevention, Preparedness, and Management of Disasters in Food and Agriculture,* Food and Agriculture Organization of the United Nations, Regional Office for Asia and the Pacific, Bangkok: 289-319.

Venkateswarlu J 1997 Sustainable crop production. In: Symposium on *Recent Advances in Management of Arid Ecosystems*. Arid Zone Research Association of India, Jodhpur

Victor US, Srivastava NN and Ramana Rao BV 1991 Moisture regime, aridity and droughts in the arid region of Andhra Pradesh. Annals of Arid Zone **30(2)** 81-91.

7

Climate Change and Drought

Uday Burman

INTRODUCTION

Climate change is real and its effects are already being experienced in several parts of the world, as is evident from the increase in average maximum temperature all over the world. The length of the growing period (LGP; defined as the number of days in any given period when there is sufficient water stored in the soil profile to support plant growth) has been reduced by 15 days in Andhra Pradesh, India leading to moisture stress to the crops and ultimately reduction in yield (Sarma and Lakshmi-Kumar 2006). Cooper et al (2009) have predicted that in the Makindu area of Kenya, average LGP might decrease from 5% to 10% by 2050, when temperatures may increase between 1°C and 2°C. Another study in western Rajasthan revealed that rise in temperature not only influenced the crop yield but also its phenology. With each degree centigrade rise in temperature, the days taken to flowering decreased by 1.5 days. Similarly, the crop matured by 2.5 days earlier with each °C rise in temperature as compared to the present maturity period of 72-75 days. At Jaisalmer seed yield of pearl millet declined by 62% due to lower rainfall and severe water stress during grain-filling stage (Singh 2008). Besides the impact of climate change on crop growth and development, LGP, water availability, etc., there is a strong possibility of further soil degradation due to loss of soil organic matter as influenced by increased temperature regimes. Increasing temperature would also increase oxidation of the organic carbon in soil. Its levels will go down further. Additionally, incidence of runoff /wind erosion may increase due to increase in extreme events. These changes may reflect themselves in poorer soil fertility, loss of soil biota, water stress and ground water depletion.

Drought is a natural, recurring feature of climate which stems from the lack of rainfall over an extended period of time (e.g. a season or several years). It

Principal Scientist,
Central Arid Zone Research Institute, Jodhpur 342003 (Rajasthan) India
Email: burmanuday821@gmail.com

occurs in high as well as low rainfall areas and is a temporary anomaly, in contrast to aridity, which is a permanent feature of the climate and is restricted to low rainfall areas. Drought is a complex phenomenon, the causes for which are variations in large-scale atmospheric circulation patterns and the locations of anticyclones, or high-pressure systems, climatic variability, sea surface temperature changes such as *El Nino*. Manmade factors e.g. deforestation, poor land and water management, green house effect aggravate the situation. However, many research analysis contest the concept of any relationship between dry climate and human activity (htpp://www.co2science.org/subject/ d/droughtasia.php).

DROUGHT MANAGEMENT

Onset of drought, its end and severity are often difficult to predict. As majority of people in India are dependent on agriculture and animal husbandry for their sustenance, drought has serious social and economic consequences. Unless drought-proofing measures are undertaken and water management is improved, we may continue to face increasing drought-proneness in coming years. Successful drought management depends on successful water management and other measures like afforestation, combating desertification and creating conditions of alternative livelihoods for people in drought affected areas. Traditional technologies used by our farmers are of immense value in combating drought at less cost.

PROJECTED IMPACTS OF CLIMATE CHANGE ON INDIAN AGRICULTURE

Agricultural sector in developing countries is particularly vulnerable to the adverse impacts of climate change. Consequences of changing climate could be one or all of the following -

- Increasing climatic variability could result in considerable seasonal/ annual fluctuations in food production.
- Droughts, hot extremes, and heat waves are known to negatively impact agricultural production. The projected increase in these events could result in greater instability in food production.
- Increasing temperature is likely to reduce fertilizer use efficiency.
- Increased fertilizer requirement to meet future food production demands.
- Greater fertilizer use leads to higher emissions of greenhouse gases.

Any adverse effect on agriculture will pose serious risks to economic growth and livelihoods across the country. Soil and water conservation technologies have been suggested as a key adaptation strategy for developing countries, in light of increased water shortages, drought, desertification, and worsening soil conditions. To combat drought and to achieve sufficient and sustainable productivity in the future, a multidisciplinary approach based on soil management, plant genomic, physiology, crop management and modelling need to be applied. Additionally, soil water conservation by optimizing land use production system and use of sustainable low cost location specific technologies need to be promoted.

Given the differences in agro-ecologies and other factors one-size-fits-all recommendations are inappropriate. Soil and water conservation technologies e.g. stone and soil bunds, grass strips, waterways, trees, and contours show positive and highly significant impacts on crop output in the low-rainfall areas, but only waterways and trees show strong and significant positive effects in high-rainfall areas. Grass strips show the largest effect on crop yields among the technologies used in low-rainfall areas. Some of the other technologies relevant in arid region are–

Managing soils: preparing for drought effects

1. Soil amendments: bentonite (to improve moisture holding capacity), sub-surface barriers (to minimize deep drainage) and mulching (to minimize evaporation losses).
2. FYM application to minimize crust formation and improve soil fertility.
3. Sowing crops on slant surface through three furrow multi crop seed cum fertilizer drill with seed pressing device useful for collecting runoff and maintain high moisture in plant root zone.
4. Shelterbelts, alternate land use to minimize evapo-transpirational losses, increase water use efficiency and maintain soil fertility.

Managing crops: insurance against aberrant weather

- Advance contingent crop planning for drought proofing.
- Planting of drought tolerant genotypes and choosing optimum plant population and row spacing.
- Planting of early and late genotypes of arid legumes in 1:1 ratio most appropriate under low as well as higher rainfall condition.
- Improved performance of mixed crop, particularly under low rainfall situation attributed to more efficient use of water.
- Adequate soil fertility for optimum plant vigour to overcome the detrimental effects of short term sporadic droughts.
- Supplemental irrigation in the event of drought at the critical stage.
- Positive additive effects of P and N nutrition along with growth regulators on plant growth under water stress condition.

For a holistic and sustained effect, research efforts in future should be directed towards the following goals

- Search for alternative crops: Diversification in favour of stress tolerant and economically more remunerative-likely to increase purchasing power.
- Improving crop breeding efficiency for developing stress tolerant genotypes by employing biotechnological tools besides utilizing selected agronomical and physiological parameters.
- Developing strategies for improving soil health for sustainable growth – precision farming.
- Continue research on impact of climate change on plants and soil.
- Continue research on standardization of agro techniques viz., crop sequence, residue management, *in situ* rain water harvesting, management of limited water and poly-house cultivation.

CONCEPTS RELATED TO DROUGHT RESISTANCE IN PLANTS

Drought resistance refers to the ability of a crop to produce its economic product with minimum loss in water-deficit environment relative to the water-constraint free management. Basically three mechanisms namely drought escape, drought avoidance and drought tolerance are involved in drought resistance.

1. **Drought escape:** It is the ability of a plant to complete its life cycle before acute soil and plant water deficit develop. This is an important strategy of matching phenological development with the period of soil moisture availability to minimize the impact of drought where the growing season is short and terminal drought predominates. This mechanism involves rapid phenological development (early flowering and maturity), developmental plasticity (variation in duration of growth period depending on the extent of water stress) and mobilization of pre-anthesis assimilates to grain.
2. **Drought avoidance:** It is the ability of plants to maintain relatively higher tissue water potentials in spite a shortage of soil moisture. Drought avoidance is achieved by maintenance of turgor through increased rooting depth, efficient root systems and internal hydraulic conductance and by reduction of water loss through reduced stomatal conductance, reduced absorption of solar radiation by leaf rolling and reduced evaporation surface i.e. leaf area reduction. Clusterbean and moth bean survive under drought conditions by doing a balancing act between maintenance of turgor and reduction of water loss.
3. **Drought tolerance:** It is the capacity of plants to withstand water deficit with low tissue water potential. Tolerance to dehydration is considered to arrive at the molecular level and depends upon membrane structure and enzyme activity. Basically it is due to the ability of the cells to maintain integrity of the cell membranes and prevent denaturation of proteins.

However, crop plants including arid legumes use more than one mechanism at a time to resist drought.

The water content and the water potential of plant tissues are the most important physiologically relevant integrators of droughts effects on plants. Once a drop in water potential develops, responses of a wide range of physiological processes are induced, either directly or by plant hormones signalling changes in water status. Literature review reveals that there is no single response pattern that is highly correlated with yield under all drought environments. Physiological traits relevant for the responses to water deficits span a wide range of vital processes. Further more, different crop developmental stages show different sensitivity to water stress in arid legumes, cereals and grasses.

Several studies have addressed yield under drought stress as a function of single physiological traits, in attempts to understand which metabolic processes or morpho-physiological traits are crucial in ensuring high yield performance under a wide range of environments. Researchers identified the following parameters- (1) Stomatal conductance in wheat cultivars, (2) Stay green in maize, rice and sorghum, (3) Osmotic adjustment (4) Selection for stem reserve contribution to grain filling. However, most of these studies relate to cereals crops like wheat, rice, maize and sorghum and these have not received adequate attention in assessment of drought tolerance in arid legumes.

POTENTIALS OF RAINFED AREAS

Having understood the impact of different field level interventions we are in a position to address to the issue of crop productivity under environmental stress to some extent. But the issue of food sustainability demands complete understanding of the role of all the factors responsible for regulating the ratio of total available food and total food demand. These factors also influence vulnerability of an ecosystem to the changes in climate. In context to India, rainfed agriculture contributing about 44% of the total food production is most vulnerable to climate variation even though it has immense potential to steer the country through the envisaged food crisis. In future, the significance of drylands in national food security will further increase due to growing population pressure and competition for land for non-agricultural uses. Rainfed agriculture will have to shoulder the maximum burden of providing food in most of the developing countries. The knowledge already exists to at least double yields in rainfed agriculture, even where water poses a particular challenge and large water investments focusing on water management are lacking. Rainfed farmers generally cannot afford the large initial investments required even for small-scale water resources management systems, despite high benefit to cost ratio and the positive impact on long-term risk reduction. Thus, it is rightly suggested that institutional reforms are required at the national level to bridge the divide in governance of water resources. Relevant departments and ministries need to be more closely connected in legal, policy and management issues (Sharma 2011).

This is more important considering the fact that rainfed agriculture is likely to play a major role in India's food security and sustainable economic growth.

In the past, smallholder farmers have coped with erratic climatic conditions by adjusting their farming practices such as early planting and replanting when crop establishment was poor. The projected increase in climate variability, however, brings new risks that will require new adaptation options. Such adaptation options need to be designed jointly with farmers to increase local relevance. Adaptation can focus on short-term operational decisions (e.g. specific timing or sequencing of farming activities), on medium-term tactical options (e.g. changes in crop rotations, or allocation of crops across fields), or on longer-term strategic decisions (e.g. to change the major crops grown, or adopt completely new activities). Tactical adaptation options include staggered planting dates on the same farm, to manage risk of drought at different times of the cropping season. Diversification of crop cultivars and soil fertility management are also potential options for adaptation (IPCC 2007). The impacts of climate change and variability on crop production are an extra pressure in addition to the existing problem of degrading soils due to continuous cropping without sufficient inputs.

DROUGHT MANAGEMENT: WORK DONE AT CAZRI

Utilizing Genotypic Variations and Agronomic Practices

Better crop management including appropriate selection of crops, their genotypes, planting dates, row spacing and plant populations, optimum fertilizer use and adoption of drought ameliorative measures may greatly help in alleviation of drought effects. Arid legumes differ in their response to drought and this helps in crop contingency planning in the event of drought (Table 1).

Table 1. Comparative reduction in yield of different arid legumes subjected to drought at different stages (Adapted from Garg, Burman & Kathju 2005)

CROPS	Per cent reduction in seed yield over control when drought experienced at	
	Vegetative stage	**Flowering stage**
Clusterbean	8.3	26.8
Moth bean	2.4	23.2
Cowpea	13.2	46.7
Mung bean	25.1	40.4

Genotypic variation in the ability of pearl millet to withstand water stress is well known since early sixties. Variation in adaptation of pearl millet cultivars to arid conditions was also indicated by relatively higher yields of land races than those of hybrids. In recent years several attempts have been made to minimize the detrimental effects of drought in arid zone crops which are briefly outlined here. Both seed size and genotypes significantly influenced early vigour, number of tillers per plant, plant height, days to wilting initiation, days to permanent wilting and dry matter production. Pearl millet plants from larger seeds took longer to initiate wilting and then permanent wilting compared with plants from smaller seeds. Further, a moderate population of 88,888 plants ha^{-1} (equal proportion of pearl millet and clusterbean) was optimum in the moisture sufficient season, and low (40,000 plants ha^{-1}) to moderate populations produced satisfactory yields in the moisture stressed season. However, it was observed that, improved performance of individual plants under wider spacing may not compensate for the losses accrued due to a decrease in plant population per unit area beyond a point (Table 2).

Table 2. Comparative performance of early and late moth genotypes under different spacing (Adapted from Burman, Garg & Kathju 2002)

Genotypes	Seed yield (kg ha^{-1})		WUE (kg ha^{-1} mm^{-1})	
	30x15	45x10	30x15	45x10
Maru Moth (L)	674	453	2.41	1.72
RMO-40 (E)	506	532	1.83	1.96

For stability of yield, planting of early and late genotypes of these legumes in 1:1 ratio was found most appropriate under low as well as higher rainfall condition. The improved performance of mixed crop, particularly under low rainfall situation could be attributed to more efficient use of water besides high photosynthetic rates (Table 3).

Table 3. Effect of planting pattern on seed yield and water use efficiency of moth bean and clusterbean genotypes (Adapted from Kathju et al 2003 and Vyas et al 2003)

Crop	Planting pattern	WUE (kg ha^{-1} mm^{-1})	
		1995	1996
Moth Bean	Early (RMO-40)	2.66	2.29
	Late (Maru moth)	1.88	2.80
	E + L	2.72	3.32
Clusterbean	Early (Suvidha)	2.89	2.66
	Late (Maru guar)	2.54	2.93
	E + L	3.20	3.16

Rainfall :1995 -266 mm; 1996 – 424 mm

Concept of life saving irrigation has also been advocated as irrigation with saline water, at the critical stage in the beginning of a dry spell, may be more rewarding than irrigation with saline waters at a belated stage when the clusterbean plants have already suffered water stress. Irrigation at belated stage may cause severe detrimental effect on crop performance due to cumulative adverse effects of drought and salinity on plant growth and leaf metabolism.

Fertility

Crops in the arid and semi-arid regions suffer not only from moisture stress but also from nutrient stress as poor fertility of arid soils is a major limitation to increasing yields. Thus, addressing the soil fertility constraints can also alleviate drought stress effects on crop productivity in arid environments (Fig 1).

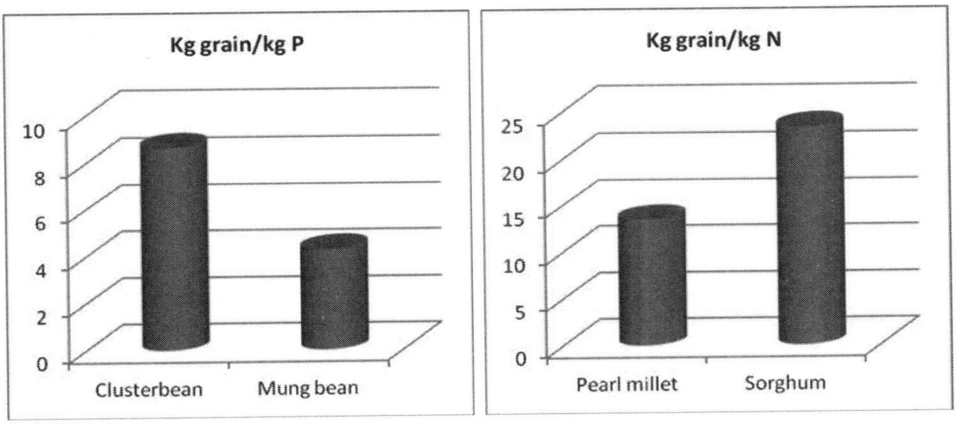

Fig. 1. Average response of P and N fertilization on seed yield of *Kharif* crops (Adapted from Narain et al 2000)

Supplemental irrigation (60 mm) during drought in pearl millet, if resources permit, coupled with N fertilization increased the grain yield from 560 to 1663 kg ha^{-1} under rainfed condition. Similar results have been obtained with sesame where supplemental irrigation to recharge 50 cm soil depth to field capacity and 60 kg ha^{-1} N application increased seed yield by approximately 33 per cent over rainfed fertilized control plants.

Studies suggested that tissue hydrature was not an infallible index of metabolic efficiency as nutritional status of plants was more critical under water deficits for enhanced leaf metabolism, photosynthesis, growth and yield. However, in case of pearl millet, response to N-fertilization was more in hybrids compared to land races (Table 4). Further, fertility mediated response in pearl millet could be observed in both sub (<300 mm) and supra (>360 mm) optimal rainfall situation (Table 5). N-fertilization at the rate of 80 kg ha^{-1}resulted in 47% increase in grain yield under low rainfall situation compared to 23% in high rainfall situation. Similar trend was observed with respect to total dry matter production, a parameter important for fodder assessment.

The adverse effect of drought, no matter at which stage of growth the water shortage was experienced, may be substantially evaded where optimum plant vigour was induced by adequate level of soil fertility. Adequate nutrition conducive for greater plant vigour brings about an efficient enzyme activity and higher chlorophyll content, despite desiccation, as compared to plants raised under low soil fertility

Table 4. Average response of N fertilization on seed yield (data mean of two years for two genotypes under each category) of pearl millet genotypes (Adapted from Garg & Burman 2011)

Category	**Grain yield (kg ha -1)**		
	N0	**N80**	**% increase**
Hybrids	1042	1595	53.0
Land races	1206	1616	33.9

High soil fertility under low moisture conditions seems to favourably influence the yields of arid legumes also. Alleviation of drought effects in arid legumes has also been achieved through P application which favourably modulates various physiological and biochemical processes Furthermore, in late moth bean genotypes where impact of drought was more severe, P application (40 kg ha^{-1}) significantly ameliorated the detrimental effects of water stress on plant water potential, relative water content, rate of net photosynthesis, levels of total chlorophyll, starch, soluble protein and nitrate reductase activity. In clusterbean also, results indicate the possibility of alleviation of water stress effects by P nutrition at least up to moderate stress level (Table 5). Interestingly, recovery upon re-watering was also more in P-fertilized than unfertilized plants.

Table 5. Effect of phosphorus nutrition on seed yield of moth bean genotypes under different moisture levels (Adapted from Garg, Burman &Kathju 2004)

Parameters	**Stress**	**P levels (kg ha $^{-1}$)**		**Percent increase by P**
		P_0	**P_{40}**	
Seed Yield*	C (-0.96MPa)	2.57	2.90	12.8
	D (-2.02MPa)	1.53	1.96	28.0

*Data mean of four genotypes (RMO-257, CZM-18, JAWALA and CAZRI MOTH-1)

K fertilization too, induced growth and yield restoration irrespective of the stage at which drought was imposed. Improved water status with K in terms of RWC and shoot water potential significantly increased the seed yield in HG-75, a drought-sensitive clusterbean cultivar subjected to water stress at flowering stage. Micronutrients too were reported to influence yields of arid legumes (Fig 2).

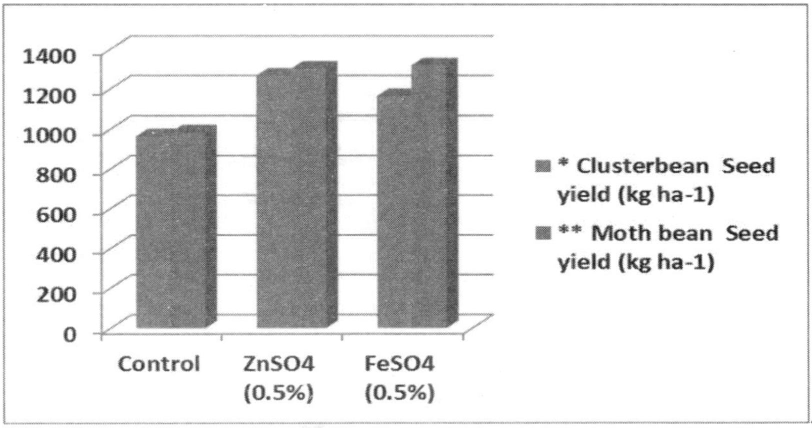

Fig. 2. Effect of zinc and iron foliar spray (at 25 & 45 DAS) on yield of clusterbean and mothbean at Bikaner (Adapted from Kumar 2005)

Growth regulators

Growth regulators have been shown to influence plant growth and performance. Few attempts of utilizing regulators for improving productivity have also been done on arid zone crops under soil moisture stress. Kinetin (5 ppm) applied as either pre-sowing seed treatment or foliar spray (at 25 and 40 DAS) or both also significantly improved the growth, dry matter production and seed yield of field grown clusterbean under moisture deficit condition. Kinetin mediated response was due to higher content of different leaf metabolites (starch, soluble protein, etc.) thereby, prolonging the active growth phase. Application of senna stem (as such or as foliar spray of its aqueous extract) also promoted growth and increased yield of pearl millet primarily due to presence of nutrients and triacantanol.

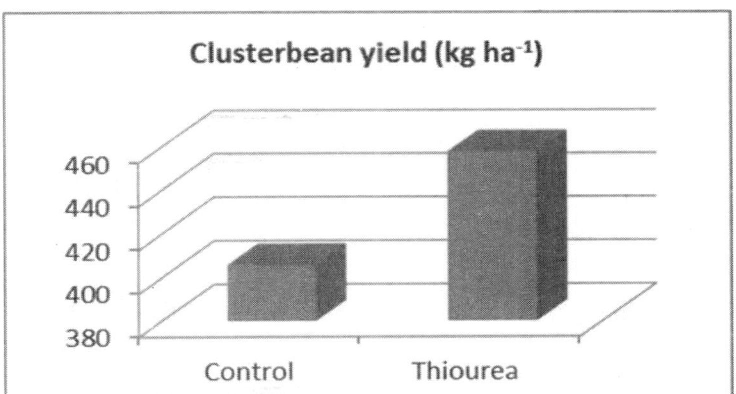

Fig. 3. Effect of thiourea on seed yield of clusterbean var. RGC-936 (Adapted from Garg *et al.*, 2006)

Similarly seed treatment with thiourea (500 ppm) followed by foliar sprays (1000 ppm) at vegetative and flowering stages significantly mitigated the detrimental effects of field drought in cluster bean through enhancement of photosynthetic efficiency and nitrogen metabolism (Fig. 3). The positive effect of thiourea was also found to be economically viable and thus remunerative for arid farmers. Adverse effects of drought could also be significantly mellowed by adequate phosphorus nutrition in combination with thiourea application due to their favourable effects on leaf area development, photosynthetic efficiency and nitrogen metabolism.

Conclusion

Thus, literature review related to climate change clearly bring out the facts that increased vulnerability (of agriculture ecosystem in the present context) may be due to lower capacity to prepare, cope and recover from climatic and non-climatic shocks and stresses. Thus, to tackle the issue in a holistic way we shall require preparedness at different levels viz.

1. Policy makers for investments on soil and water conservation technologies.
2. State administration for implementation through convergence of schemes.
3. Agricultural scientists for crop diversification, development of stress tolerant genotypes, continued research on impacts of climate change and precision farming.

REFERENCES

Burman U, Garg BK and Kathju S 2002 Effect of spacing on seed yield and physiological traits in moth bean (*Vigna aconitifolia*), *Indian Journal of Agricultural Sciences* **72** 76-79

Cooper P, Rao KPC, Singh P, Dime J, Traore PS, Dixit P, and Twomlow SJ 2009. Farming with current and future climate risk: Advancing a 'Hypothesis of Hope' for rainfed agriculture in the semi-arid tropics. *Journal of SAT Agricultural Research* **7** 1-19

Garg BK and Burman U 2011 Physiological basis of yield improvement of arid zone crops through nutrient management, *Plant Stress (Special Volume on Abiotic Stress)* 73-81

Garg BK, Burman U and Kathju S 2004 The influence of phosphorus nutrition on the physiological response of moth bean genotypes to drought, *Journal of Plant Nutrition and Soil Science* **167** 503-508

Garg BK, Burman U and Kathju S 2005 Comparative water relations, photosynthesis and nitrogen metabolism of arid legumes under water stress, *Journal of Plant Biology,* **32** 83-93

Garg BK, Burman U and Kathju S 2006 Influence of thiourea on photosynthesis, nitrogen metabolism and yield of clusterbean (*Cyamopsis tetragonoloba* (L.) Taub.) under rainfed conditions of Indian arid zone, *Plant Growth Regulation* **48** 237-245

IPCC 2007 Climate change 2007 the physical science basis. In: Solomon, SD, et al. (Eds.), Contribution of Working Group I to the Fourth Assessment Report of the Intergovernmental Panel on Climate Change. Cambridge University Press, Cambridge, United Kingdom and New York, NY, USA, 996 Pages

Kathju S, Garg BK, Vyas SP and Lahiri AN 2003 Sustainable production of moth bean through genotype management under arid environments, *Journal of Arid Environments* **53** 137-143

Kumar D 2005 Research Achievements of AICRP on Arid Legumes.

Narain P, Sharma KD, Rao AS, Singh DV, Mathur BK and Ahuja UR 2000 *Strategy to Combat Drought and Famine in the Indian Arid Zone*. CAZRI, Jodhpur, 1-65p.

Sarma AALN and Lakshmi Kumar TV 2006 Studies on crop growing period and NDVI in relation to water balance components *Indian Journal of Radio and Space Physics* **35** 424-434

Sharma KD 2011 Rainfed agriculture could meet the challenges of food security in India, *Current Science* **100** 1615-1616

Singh DV 2008 Effect of climate change on pearl millet productivity in Thar Desert. *DEN News* **40** 2-3

Vyas SP, Garg BK, Kathju S and Lahiri AN 2003 Genotypic management of clusterbean for sustainable production under arid environment, *Indian Journal of Pulse Research* **16** 104-107.

8

Agro Meteorological Observatory

H.M. Meena

INTRODUCTION

Agrometeorology, abbreviated from agricultural meteorology, puts the science of meteorology to the service of agriculture, in its various forms and facets, to help with the sensible use of land, to accelerate the production of food, and to avoid the irreversible abuse of land resources (Smith 1970). The science of agro meteorology has great practical utility in protection against, or avoidance of adverse climate risks (crop losses due to drought or floods). Weather elements which influence agricultural operations and crop production can be forecast for different time spans; hailstorm, tornados, and flash flood can be forecast up to 12 hours in advance. Based on the forecasts of these elements, decisions can be made in advance in agricultural operations and planning so as to make the best use of favourable weather conditions. However, agrometeorological information can be used in efficient land use planning; determining suitable crops for a region; risk analysis of climate hazards and profit calculations in farming; production and harvest forecasts; and in adoption of farming methods and choice of farm machinery. Agrometeorological observatory is a place where all the necessary instruments are maintained to observe and record various weather parameters at specific time interval. Each agrometeorological station should be sited in a place which is representative of natural conditions and agricultural characteristics of the region in which it is located (Bernardi 1983).

CLASSIFICATION OF AGROMET OBSERVATORIES

Depending on the availability of instruments and number of weather elements observed agrometeorological observatories are classified into three major (A, B and C) and three minor (D, E and F) classes (Sahu et al 2012).

Scientist (Agril. Meteorology),
Central Arid Zone Research Institute, Jodhpur-342003
Email:hmmeena82@gmail.com

Class A Observatory: These are provided with eye reading instruments as well as self-recording instruments. The observations are recorded every hour round the clock.

Class B Observatory: Instruments are similar to Class A observatory but observations are made at least twice a day.

Class C Observatory: Instruments are similar to Class A observatory but observations are recorded only once a day.

Class D, E and F observatories: These have less number of instruments or are non-instrumentals.

Fig. 1. Agrometeorological observatory at CAZRI Jodhpur

SELECTION OF SITE FOR AGROMETEOROLOGICAL OBSERVATORY

The Indian Meteorological Department (IMD) which is the principal government agency in all matters pertaining to meteorology in India, recommends the following procedure in the selection of site for meteorological observatory.

- Site should be well exposed, bare, levelled plot taking care of the proposed buildings, roads, canals, etc., which affect the exposure.
- When the observatory is to be located in an agricultural farm, it should be as far as possible so chosen that it is the representative of principal agricultural soils of the area.
- The site should be free from water logging, high structures, tall trees and hills, especially on the east and west as these cast shadows on the instruments. The above obstructions should be away by at least ten times their height. Where radiation or sunshine observations are envisaged, there should be no substantial objects to the east or west subtending an

elevation angle of more than 3 degrees with respect to the level of the radiation instrument or sunshine recorder.

- The size of the observatory has to be decided depending upon the number of instruments to be installed. The general recommendation for size of the plot is 54 m N-S x 36 m E-W for an agro-meteorological observatory

TIME OF OBSERVATIONS

As meteorological elements vary with time, it is necessary that they should be recorded at a particular time on every occasion. In India, the main observations are recorded as per the guide lines of IMD. at 08:30 and 17:30 Indian Standard Time (IST). At agromet observatories, the observations (wind speed and direction, soil temperature, maximum and minimum temperature, wet bulb and dry bulb temperature) are recorded at 07:00 and 14:00 hours LMT (Local Mean Time) except evaporation and rain fall which are recorded at 08:30 hours IST. Thermometer maximum is set on 7.00 LMT and minimum on 14.00 LMT. Dew is measured before sunrise during dew season, sunshine card is changed before sunrise or after sunset. The observation of the various parameters of weather in agromet observatory is as follows (Table 1).

Table 1. Instruments and observations in the agrometeorology observatory

Parameters	Instruments	Unit	Observation Frequency
Rainfall	Ordinary rain gauge, Self-recording rain gauge	Millimetres (mm)	Twice daily 8.30 a.m. 2.30 p.m. continuous (The time of placing the chart 8.30 a.m.)
Air Temperature	Dry and Wet bulb thermometers, Min. & Max. thermometer Thermograph	Degree Celsius (°C)	Twice daily 7 a.m. & 2 p.m. The time of removing or placing the chart is 8.30 a.m.
Sun shine duration	Sun shine recorder	Hours (h)	Continuous- The time for placing the sun shine card is before sunrise
Evaporation	U.S. Class A Evaporimeter	Millimetre (mm)	Twice daily 8.30 a.m. & 2.30 PM
Relative humidity	Dry and Wet bulb thermometers, Assman-Psychrometer Hygrograph	Per cent (%)	Twice daily 7 a.m. & 2 p.m., Continuous and time for placing the chart 8.30 a.m.
Soil temperature	Soil thermometer	Degree Celsius (°C)	Twice daily 7 a.m. & 2 p.m.
Wind speed	Cup Anemometer	Kilometre per hour (kmph)/ meter per second (mps)/ knots (kts)	Twice daily 7 a.m. & 2 p.m.
Wind Direction	Wind wane	Degree clock wise from north on the scale/00-36 where 00 & 36 is the wind from where the north, 09 from the east and 00 refers calm	Twice daily 7 a.m. & 2 p.m.

Cloud amount	Oktas (1/8 of the celestial dome)	Visual observed in the observatory	At the time of observation
Dew	Dew Gauge	Millimetre (mm)	One time during winter season before sunrise (install after sunset of previous day)

(Srivastava and Tyagi 2015)

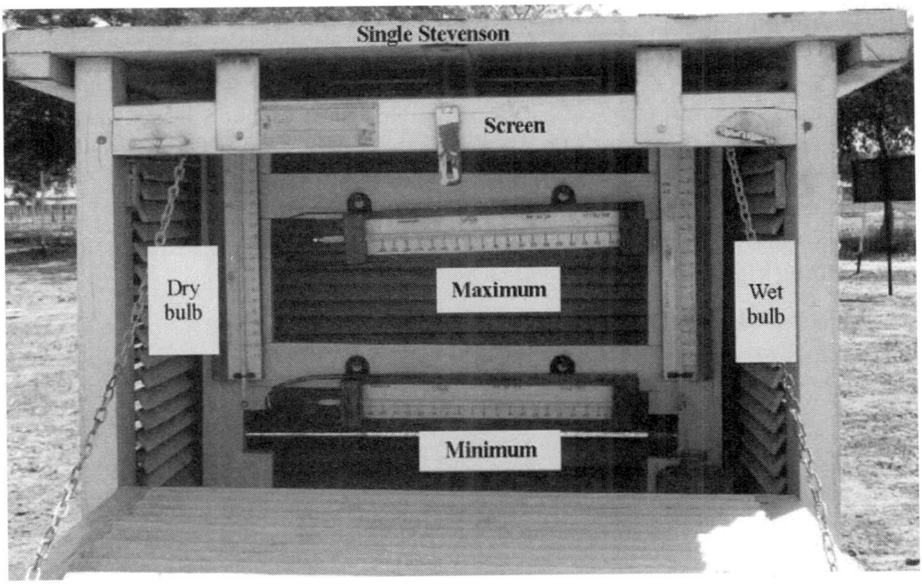

Fig. 2. Single stevenson screen

INSTRUMENTS FOR MEASURING WEATHER PARAMETERS

Stevenson screen: It is a wooden box in which maximum and minimum thermometers are placed inside the screen horizontally whereas dry bulb and wet bulb are hanged vertically (Fig. 2). It is installed at 1.22 meter height from ground surface and provides all essential conditions for the exposure of the thermometers like free access of air to bulb of the thermometers and protect the thermometers from the direct exposure to sunrays and raindrops. The maximum thermometer is a mercury thermometer with a constriction in the bore near the bulb and it measures maximum temperature in a day. Minimum thermometer indicates lowest temperature of day and it is an alcoholic thermometer having a light narrow index. Only alcoholic thermometer is used for measuring minimum temperature because alcohol freezes at -130°C whereas mercury freezes at -39°C. The screen is having two other thermometers also which are placed vertically and are used to work out the relative humidity, dew point temperature and vapour pressure. One is mounted a little lower than the other and has its bulb covered with a piece of muslin or wicking that can be wetted for observation. It is wet bulb thermometer. When the instrument is swung freely in the air or is aerated by a fan, the loss of heat required to evaporate water from the wet-bulb will cause it to show a lower temperature reading. The difference between the readings is called the wet-bulb depression. When the dry-bulb temperature, the wet bulb depression, and the atmospheric pressure are known, any of the standard expressions of humidity can be determined from a series of hygrometric tables.

Soil thermometers

These are mercury in glass type thermometers which are used for recording soil temperature at desired depths. The stem of the thermometers is bent at 120° just near the bulb. These thermometers are held by iron stands in inclined position making 60° angle with soil surface. Soil temperature plays a major role in the movement of water and nutrients from the soil. The surface of soil is warmer than the air above, so that surface thermometer is used for measuring surface soil temperature (Fig. 3).

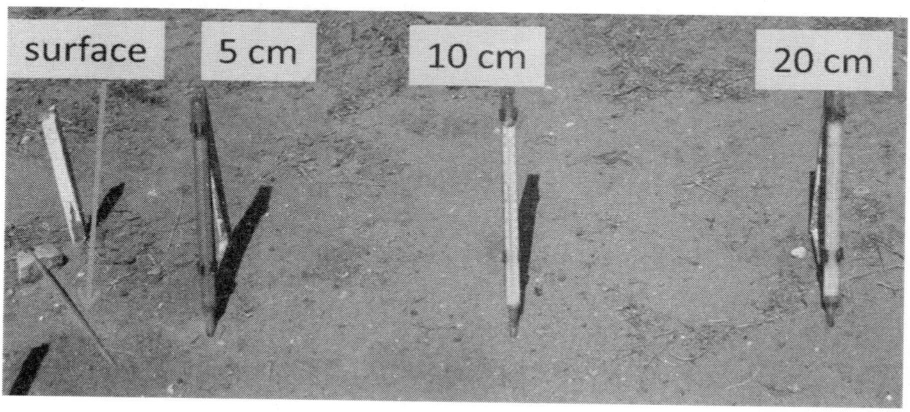

Fig. 3. Soil thermometers at various soil depths.

US Class A Open Pan Evaporimeter

It is a device that measures the amount of water evaporated in certain time from a surface of water. It has a cylindrical reservoir of 120 cm diameter and 25 cm height (Fig. 4a). The reservoir rests on a wooden platform (120 cm x 120 cm) placed on ground. It is covered with wire mesh for avoiding of water loss by birds. Water loss by evaporation in a day is measured by accurately adding water with help of measuring cylinder up to tip of the point in the stilling well. It is measured at 8.30 a.m.

Ordinary rain gauge

Ordinary rain gauge is a device used for measuring the amount of rainfall in a day (mm or cm or inch). It is installed at 30 cm height above ground (Fig. 4b). It comprises a funnel with five inches diameter in the mouth, a container to which the funnel is shouldered and finally a metallic cylinder that holds both funnel and container. A measuring flask is used for the measurement of amount of rain water received by container through the funnel. It is also measured at 8.30 a.m. Now-a-days, sensor or self-recording rain gauge or automatic weather station can measure rainfall hourly or any time interval.

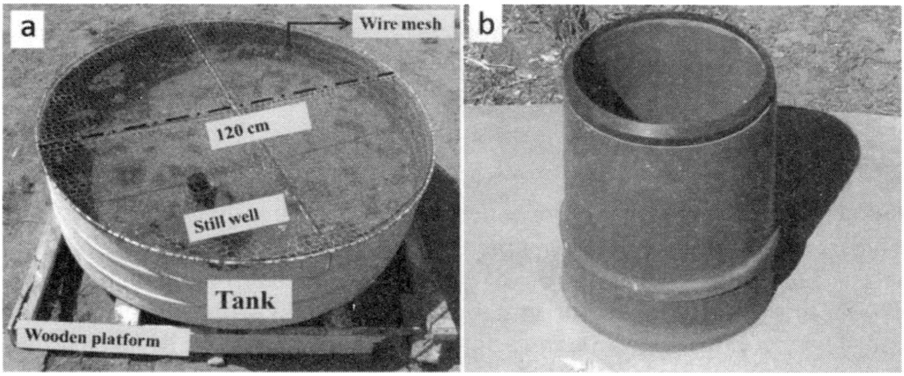

Fig. 4. Class A open pan evaporation (a) and ordinary rain gauge (b)

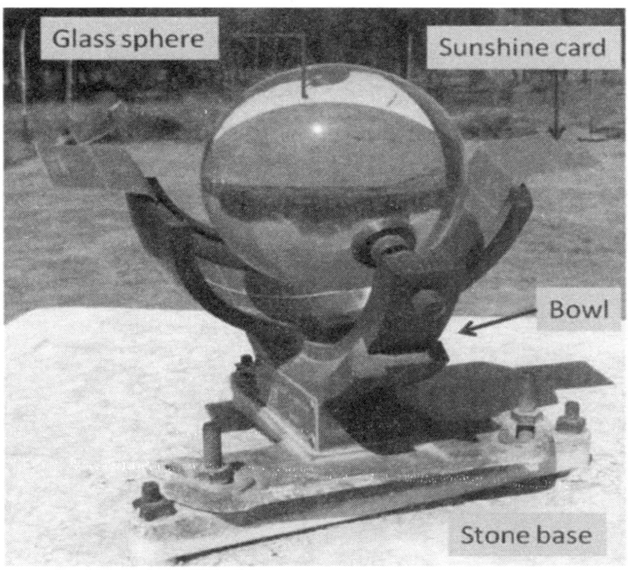

Fig. 5. Campbell stock sunshine recorder

Bright Sunshine Recorder

The sunshine recorder is installed on a masonry pillar of 3.04 meter height from the ground. It works on the principle of convex lens. A convex lens formed by a spherical mounted sphere focuses the sun rays coming parallel to its principle axis at a distance equal to its principle focus. When the sun rays fall on the glass, a strip of prepared paper held in a grove at the focus length starts burning. The card is made of celluloid with specific width and length. When the sun does not shine, the burn point of the paper the sun light hours (duration) of the sun shine can be recorded. During the year, we use three types of sunshine cards according to the season. During winter season (13th Oct -28th Feb) short curve card is used in the upper groove and from 13th April to 31st August long curve sunshine card is used in the lower groove, and for other seasons straight card is used (Fig. 5) during the period of 3rd Sept- 14th Oct and 1st March-11th April.

Wind vane

Wind vane is installed over a wooden plank at 3.08 m height and fixed on a wooden post (Fig. 6). It is an instrument that indicates the direction of wind. It consists essentially of a broad arrow head placed in ball bearings to enable the arrow to move freely in the horizontal plane. The arrow indicates the direction of wind. There are two ways of expressing wind direction that are by sides as North, South, etc., and by degrees as 360°, 110° etc.

Cup Anemometer

Anemometers are instruments used for the measurement of wind velocity. The most common type anemometer is the Robinson Cup Anemometer. Either three or four metal hemispheres are mounted on arms which rotate freely about a vertical axis (Fig. 6). The vertical spindle, about which the cup rotate is connected to a mechanical counter called cyclometer through a gear system from which the number of rotations are counted and converted into speed calibrated in km/hr. Digital anemometer gives more precise and accurate reading.

Fig. 6. Cup anemometer and wind vane

Assman Pyschrometer

It is used to measure air temperature and relative humidity under crop field. It basically consists of two thermometers, one wet bulb and one dry bulb thermometer, and a fan. The fan is at the top. The bulbs of the two thermometers are covered by metal sleeves whose outer surface is coated with chromium (Fig. 7). These are thermally insulated from each other through a central tube and has a range from 0 to 60°C. When the fan is kept on, the air through inner sleeves (metal sleeves provided at the bulb of each thermometer) and annular spaces enters the apparatus. The air travels through the central tube and goes out of the apparatus through the slits provided at the top near the fan. The fan is kept on for an interval of approximately 3 minutes.

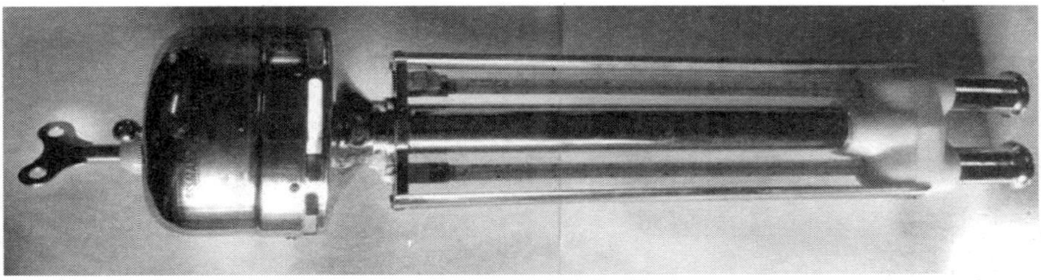

Fig. 7. Assman pyschrometer.

Leaf wetness sensor

The leaf wetness periods are recorded using leaf wetness sensor. The sensor is essentially an electronic leaf. The output from sensor is in mV. This output is drastically reduced if there are any water/rain drops on the sensor. These events are counted at short intervals of few minutes to get the duration for which the leaf was wet. The wetness period is converted into some index developed under control/laboratory conditions for disease incidence versus threshold index value to advise on weather dependence spray schedule for effective and economic control disease (Bapuji Rao and Rao 2012).

Soil heat flux

The flux of heat into and out of soil can be measured directly by 'thermal flux plates". Soil heat flux plate may be either disk or wafer shaped. Designers try to keep the size of the plates small so as to minimize the distortion in the soil temperature field and to minimize trapping that these impermeable object must cause when emplaced in soil (Rosenberg, 1974). A typical model of a soil flux place is shown in Fig. 8.

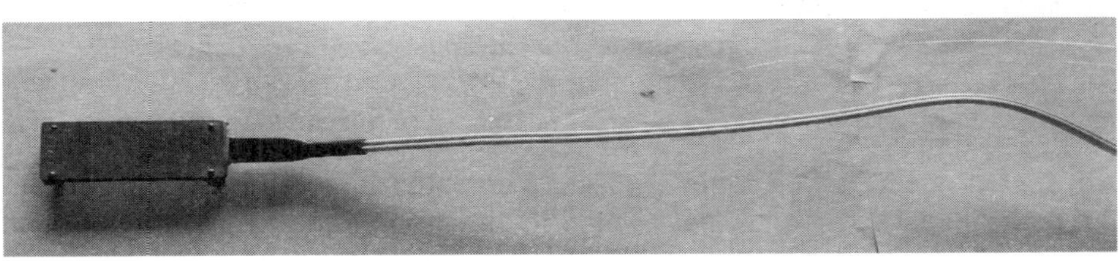

Fig. 8. Soil heat flux plate

Mini-Lysimeter

Mini-lysimeters are generally used to quantify accurate and minor changes in amount of soil moisture content, drainage water, evapotranspiration, and other water balance components within a soil monolith.

Fig. 9. Mini-lysimeter

Fig. 10. Automatic weather station at CAZRI Jodhpur

AGRO-AUTOMATIC WEATHER STATION (AWS)

The AWS consists of a data logger, set of sensors, power supply, solar panel, mounting stand and other accessories (Fig. 10). The sensors are installed for measurement of wind direction and speed, soil and air temperature, global radiation, leaf wetness and leaf temperature, soil moisture and rainfall. Global radiations are measured with pyranometer. Rainfall is measured with tipping bucket rain gauge. AWS should be located in such a place that it represents the general agro climatic conditions of the watershed area. The AWS records weather data on hourly basis.

REFERENCES

Bapuji Rao B and Rao VUM 2012 *Crop-weather dynamics,* Central Research Institute for Dryland Agriculture, Hyderabad, 350p.

Bernardi M 1983 Manual of meteorological instruments and meteorological and phonological observations. C.I.L.S.S., Niger.

Rosenberg JN 1974 *Microclimate: the biological Environment*. New York: John Willey & Sons

Sahu DD, Chopada MC and Kacha HL 2012 *Practical agrometeorology*. Agrobios (India), Jodhpur.

Smith LP 1970 Aims and extent of agricultural meteorology. *Agricultural Meteorology* **7** 193-196

Srivastava AK and Tyagi PK 2015 *Practical Agricultural Meteorology*, New India Publishing Agency (India).

WMO 2010 *Guide to agricultural meteorological practice*, WMO vol. 134. http://www.wmo.int/pages/prog/wcp/agm/gamp/documents/WMO_No134_en.pdf

9

Soil Management Strategies to Improve Moisture Availability

D.K. Painuli

INTRODUCTION

Scarcity of soil moisture is the major limiting factor to crop production under rain dependent agriculture. This might be due to low and erratic rainfall but more so due to inappropriate moisture management. Soil moisture management can play a major role in mitigating the impact of drought by conserving moisture in the soil and providing at least some production. Increasing infiltration and retention of water in soil profile (storage), and minimising runoff and evaporation are therefore, the critical pathways to improve soil moisture content of the rooting zone. The soil management strategies to achieve the objective of improved moisture availability depend on various edapho-climatic situations.

SOIL STRUCTURE AND ITS RELEVANCE TO SOIL MOISTURE PROCESSES

Soil structure refers to the size, shape and arrangement of solids and voids, and the continuity of pores and voids. Columnar, Blocky, Granular and Platy are the four basic structural types of soil. Water percolation is rapid in granular, slow in platy, and moderate in other types of structures. For optimum capture, percolation, storage and use of soil moisture, followings features are important:

- Capacity to allow water entry into profile (infiltration) and resist runoff.
- Capacity to allow water to move readily through the profile (permeability).
- Capacity to store the acquired water in the rooting zone (water holding capacity) and release it to plant roots when required.

Principal Scientist
Central Arid Zone Research Institute, Jodhpur 342003 (Rajasthan) India
Email: painulidk@gmail.com

- Capacity to resist water vapour loss from soil surface to atmosphere (evaporation).

The above soil structure dependent physical processes and their capacities, in general, depend on soil texture, depth, organic matter content, biological activities and chemical composition. The number, size and connectivity of pore spaces play crucial role in determining the amount of water that a soil can absorb, hold and supply. Pores of different size ranges play different roles in soil moisture retention (Table 1).

Table 1. Functions and sizes of soil pores (Hamblin 1985)

Pore size (mm diameter)	Description of pores	Functions of pores
< 0.0002	Residual	Retain water that plants cannot use
0.0002-0.05	Storage	Retain water that plants can use
> 0.05	Transmission	Allow water to drain out and air to enter
> 0.1 to 0.3	Rooting	Allow crop roots to penetrate freely
0.5-3.5	Worm holes	Allow water to drain out and air to enter
2-50	Ant nests and channels	Allow water to drain out and air to enter

Porous surface with elongated (continuous) pores in the size range 0.0002-0.05 mm diameter is desirable for infiltration and retention of water. However, there is no such general agreement on desired surface structure attributes viz. size of voids and aggregates for minimizing evaporation. Accelerated soil drying in the presence of 4-8 and 8-16 mm diameters voids has been reported by Ojeniyi and Dexter (1984). Based on literature review, Heinonen (1979) concluded a homogeneous layer of graded aggregates sized in 0.5 to 2.0 mm range provided the most efficient evaporation control. He explained, in the 0.5 to 2.0 mm range both capillary flow and turbulent gas flow are small. If aggregate size decreases below 0.5 mm, capillary flow increases and water is lost through increased evaporation due to proximity of water to the soil surface. If aggregates are over 2.0 mm turbulent evaporative moisture loss increases. However, practical problems arise with the creation of ideal evaporation control layers. Areas where moisture is the limiting factor in crop production are normally associated with high winds and soil erosion problems. Chepil (1955) stated that dry soil particles 0.84 mm in equivalent diameter and smaller are erodible. Thus, while the ideal aggregate size range of 0.5 to 2.0 mm diameter suggested by Heinonen (1979) may reduce evaporative soil moisture losses the adverse effect of soil wind erosion may eliminate any gain in crop yield.

SOIL COVER AND ITS RELEVANCE TO SOIL MOISTURE PROCESSES

Soil cover improves infiltration by preventing damage to surface structure and reducing runoff. It reduces evaporation by:

- reducing energy reaching the soil surface by reflecting or intercepting the solar radiation and therefore keeping vapour pressure low at the soil surface; and
- Reducing vapour diffusion from the site of evaporation to the atmosphere by increasing the thickness of non-turbulent air above the soil surface.

Under field conditions direct and precise measurement of evaporation from residue covered soil is difficult because of the interacting factors such as residue orientation (flat or standing, which affects porosity and thickness of the layer), non-uniformity of the residue layer, rainfall interception by residue and the aerodynamic roughness. The reduction of evaporation from soil with crop residues is however, reflected by increased profile water storage during fallow. In a field study Parihar et al (1979) observed that freshly cut *basooti* (*Promna mucronata* L.) weed mulch applied @ 10 Mg ha^{-1} a month prior to harvest in standing corn affected the water storage favourably in three out of seven years. It increased the 0-180 cm profile storage by 20 mm or more at harvest of the crop. Critical analysis of rainfall distribution pattern after mulch application indicated that in years when the profile was wetted before or immediately after mulching and there were no rains till harvest of maize, the moisture conservation effect was absent. But when rains were received in the later period after mulching but prior to maize harvest, moisture storage with mulch was significantly higher than without it. However, even in years of negligible effects on total moisture storage, mulching tended to keep the surface layers more moist for a longer period. Similarly, Jalota and Prihar (1979) reported that the effect of residue mulch varied with rainfall characteristics and evaporativity (E_0, evaporativity is defined as the maximum flux at which atmosphere can vaporize water from free water surface). When rain showers were heavy, residue mulch was more efficient under moderate than under low E_0 because in the latter case soil surface does not dry out quickly. Under high E_0 and light rain showers residue was less efficient as most of the rain water was lost from the mulch itself. Mulching of a sandy loam soil with 6 Mg ha^{-1} wheat straw during high E_0 period of March to June (82 days) reduced evaporation loss from 123 to 101 mm. During the low E_0 months of September to January (123 days) mulching reduced water loss from 58 mm to 54 mm. The rainfall during the high and low E_0 was 119 mm and 56 mm respectively, which occurred in 15 and 4 storms. Thus, mulching with crop residues reduced evaporation only where the soil surface was maintained at high moisture content by frequent rains or a high water table.

SOIL MANAGEMENT STRATEGIES TO IMPROVE MOISTURE AVAILABILITY

Soil management to enhance moisture availability involves tillage, amendment, and soil cover.

Tillage based strategies

- Making surface soil porous (open).
- Making surface cloddy (rough).
- Breaking crust/ compact surface layer.
- Homogenising texturally different (layered) soil profile.
- Increasing soil depth by breaking sub-surface pan or compact zone.
- Modifying land configuration e.g. ridge-furrow system.
- Field bunding to improve infiltration (reduce runoff).
- Breaking capillary connectivity.
- Creating dust mulch.
- Creating cloddy surface to reduce evaporation.

Amendment based strategies

- Application of organic matter e.g. FYM, manure, compost, industrial wastes in soil.
- Application of tank silt in coarse texture soil.
- Application of gypsum, industrial wastes in sodic soil.
- Application of fertilizers.

Soil cover based strategies

- Residues retention.
- Application of grasses and weeds as mulch.
- Application of tree leaves and tender twigs.

Rapid degradation of soil under prevalent agriculture practices (mechanization, use of chemicals etc.), climate change, and socio-economic demands are some of the compelling factors which in recent times have driven agriculture to adopt eco-friendly practices. One important objective is to enhance soil carbon stocks, which is universally acknowledged to improve soil structure (Fig. 1) and therefore moisture availability. The strategies preferably to be followed in integrated manners are:

- Reduced soil disturbance including controlled traffic.
- Increased input of organic matter.
- Maintaining soil cover.
- Growing resource compatible crop/cropping system including leguminous crop.
- Balanced fertilizer application.

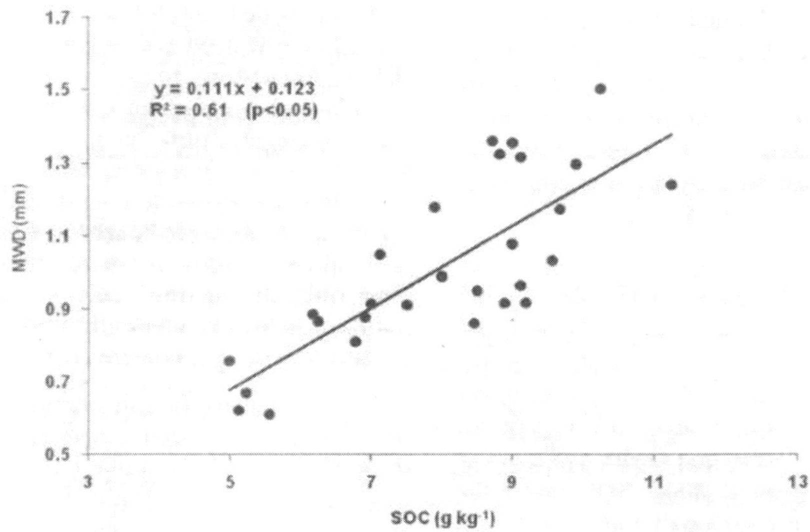

Fig.1. Soil organic carbon (SOC) and mean weight diameter (MWD) relationship in a vertisol of central India (Mohanty et al 2012)

ILLUSTRATIVE EXAMPLES TO EXHIBIT IMPROVED MOISTURE AVAILABILITY BY SOIL MANAGEMENT STRATEGIES

Followings are the three examples from three different agro-climatic locations to exhibit effective application of soil management strategies to improve moisture availability.

Example 1: Reduced tillage and mulch (Palampur, Western Himalayan Region)

In the hills of north-west India rainfed wheat in maize-wheat cropping system suffers from lack of optimum moisture at sowing. Acharya et al (1998) reported data of a 3 years field experiment on a silty clay loam at Palampur, mean annual rainfall 2500 mm, conducted to evaluate the effectiveness of mulch applied in standing maize in combination with conservation minimum tillage in mitigating water scarcity problem that is encountered in wheat. These workers reported improved soil organic carbon level and higher earthworm population in mulched treatment and consequent higher profile moisture (Table 2) and yield of wheat. Practice of conservation tillage with mulch is recommended over conventional tillage with mulch or conventional farmer practice.

Table 2. Soil water content (m^3 m^{-3}) at two stages of wheat growth during water scarcity periods at Palampur (Acharya et al 1998)

Treatment	Sowing						30 days after sowing					
	0-15 cm depth			15-30 m depth			0-15 cm depth			15-30 m depth		
	1989-90	1990-91	1991-92	1989-90	1990-91	1991-92	1989-90	1990-91	1991-92	1989-90	1990-91	1991-92
T_1	0.23	0.24	0.19	0.36	0.24	0.26	0.27	0.13	0.11	0.31	0.15	0.16
T_2	0.24	0.24	0.21	0.37	0.24	0.27	0.33	0.15	0.15	0.37	0.19	0.21
T_3	0.13	0.17	0.10	0.32	0.20	0.20	0.25	0.12	0.09	0.30	0.13	0.16

T_1-Mulching previous standing maize with tender twigs of wild sage (lantana) @ 10 Mg ha^{-1} (dry weight) towards the end of August when monsoon rains started receding followed by its incorporation with two or three diggings with a spade to 0.15 m depth (conventional tillage) immediately prior to sowing of wheat in 1st week of November.

T_2- Mulching previous standing maize with tender twigs of wild sage (lantana) @ 10 Mg ha^{-1} (dry weight) towards the end of August when monsoon rains started receding followed by manual removal of weeds and sowing wheat in 1st week of November in a furrow made with a hand driven plough without any preparatory tillage (conservation tillage) and retaining the residues of wild sage as mulch.

T_3- Two or three diggings with a spade to 0.15 m depth (conventional farmer practice of soil tillage) immediately prior to sowing of wheat in 1st week of November.

Example 2: Reduced tillage and residue retention, Ballowal Saunkhri, Trans-Gangetic Plains Region

In "Khandi" area of Punjab in spite of productive alluvial soil, adequate annual rainfall (750-1150 mm) and favourable climatic conditions yields of rainfed maize and wheat in maize – wheat system are low. Based on observations recorded after the fifth crop of wheat in a five years study at Ballowal Saunkhri, Ghuman and Sur (2001) reported higher cumulative infiltration (Fig. 2), higher water retention (Table 3), higher organic carbon, higher aggregation, and higher water use efficiency for minimum tillage treatment (drilling small holes with a hoe for maize seed placement, or trenches with a manual plough for wheat seed placement) plus residue retention (residue of previous crop was left on the surface as mulch @ 3 Mg ha^{-1}) compared to minimum tillage and conventional tillage (two disking to about 0.1 m depth followed by one ploughing i.e. levelling with a 3 m long wooden bar before seeding) treatments. These researchers attributed the observed improvement in water availability in minimum tillage plus residue treatment to improved soil carbon level and the consequent improvement in soil structure.

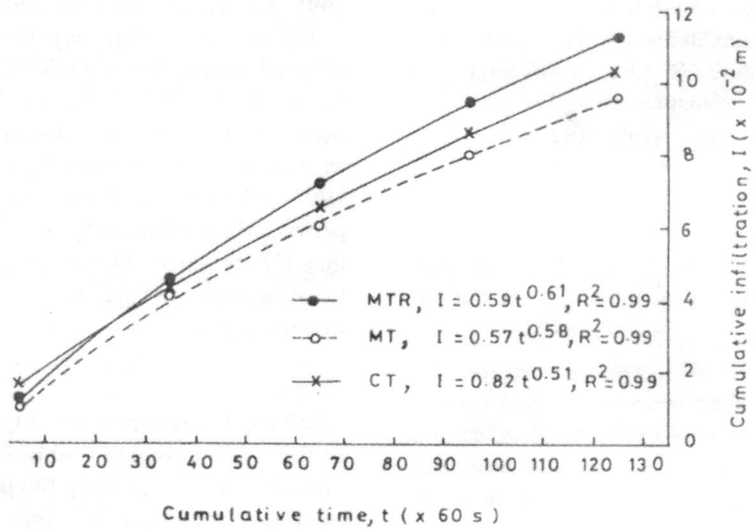

Fig. 2. Effect of tillage and residue mulch on cumulative infiltration of water into a field soil at the end of fifth maize - wheat rotation. MTR: minimum tillage + residue; MT: minimum tillage without residue; CT: conventional tillage without residue (Ghuman and Sur 2001).

Table 3. Effect of tillage and crop residue mulch on soil water retention in 0-5 cm depth at different suctions (Ghuman and Sur 2001)

Treatment	Water content (g kg^{-1}) at suction (kPa)			
	0	**5**	**10**	**20**
MTR	287	207	153	120
MT	246	160	127	106
CT	260	180	133	113
$LSD_{0.05}$	20	20	13	NS

MTR- Minimum tillage (drilling small holes with a hoe for maize seed placement, or trenches with a manual plough for wheat seed placement) plus residue retention (residue of previous crop left on the surface as mulch @ 3 Mg ha^{-1}).

MT- Minimum tillage (drilling small holes with a hoe for maize seed placement, or trenches with a manual plough for wheat seed placement).

CT- Conventional tillage (two disking to about 0.1 m depth followed by one ploughing i.e. leveling with a 3 m long wooden bar before seeding).

Example 3: Compost application, Jodhpur, Western Dry Region

In a field study on a sandy loam soil (Pal series) at CAZRI, Jodhpur 0, 2.5 and 5.0 t compost ha^{-1} yr^{-1} with varied quantities of neem cake primarily to control termites were applied to Pearlmillet (*Pennisetum glaucum*) - Cluster bean

(*Cyamopsis tetragonoloba*) and Cluster bean - Pearlmillet crop rotations for six years. At the end of 6th crop cycle various soil properties including soil carbon, field capacity moisture, hydraulic conductivity and infiltration rate were measured. These revealed improvement with compost application and therefore improved soil moisture condition (Table 4, unpublished data of D K Painuli and A K Sharma). As a consequence, crops were able to withstand seasonal drought and produced higher yields (Table 5, Personal communication, A K Sharma). Therefore, compost application is recommended for improved use of rainwater.

Table 4. Influence of organic inputs on physical properties of Pal series soil at crop harvest after the 6th cycle of pearl millet - cluster bean and cluster bean - pearl millet rotations. Each value is a geometric mean of six values pooled over two rotations (unpublished data of D K Painuli and A K Sharma)

Compost+*Neem* cake (mg ha^{-1} yr^{-1})	Bulk density (g cm^{-3})	Cone index (kg cm^{-2})	Hydraulic conductivity (cm hr^{-1})	Field capacity moisture (% by weight)	Organic carbon (%)	Infiltration rate (cm hr^{-1})
			0-10cm		0-15cm	
0.0+0.0	1.67	5.10	3.34	7.87	0.15	7.8
2.5+0.2	1.65	4.76	3.54	8.18	0.17	9.8
5.0+0.4	1.61	3.91	5.44	8.31	0.21	11.5
Sed	0.03	ns	0.61	0.40	0.03	1.2

Table 5. Influence of organic inputs on yield of pearl millet and cluster bean in the 6th year (unpublished data of A K Sharma)

Compost+*Neem* cake (mg ha^{-1} yr^{-1})	Pearl millet (kg ha^{-1})	Cluster bean (kg ha^{-1})
0.0+0.0	393.6	222.8
2.5+0.2	763.6	483.6
5.0+0.4	812.5	511.2

CONCLUSION

Land management practices which increase soil carbon stocks such as conservation tillage with residue retention, compost addition etc. improve soil structure and as a result enhanced capacity of the soil to capture and supply to crop more of rainwater. This enhanced capacity of soil to capture and supply moisture enables crop to withstand shortage of water including drought and produce higher yield. Therefore, management practices which increase soil carbon stocks should be encouraged in general but more vigorously in rainfed regions to cope with low and erratic rainfall including the drought.

REFERENCES

Acharya CL, Kapur OC and Dixit SP 1998 Moisture conservation for rainfed wheat production with alternative mulches and conservation tillage in the hills of north-west India. *Soil and Tillage Research* **46** 153-163

Chepil WS 1955 Factors that influence clod structure and erodibility of soil by wind. V. organic matter at various stages of decomposition. *Soil Science* **80** 413-421

Ghuman BS and Sur HS 2001 Tillage and residue management effects on soil properties and yields of rainfed maize and wheat in a subhumid subtropical climate. *Soil and Tillage Research* **58** 1-10

Hamblin AP 1985 The influence of soil structure on water movement, crop root growth, and water uptake. *Advances in Agronomy* **38** 95–158

Heinonen R 1979 Soil management and crop water supply. Swedish University of Agricultural sciences, Uppsala, 106 page

Jalota SK and Prihar SS 1979 Soil water storage and weed growth as affected by shallow tillage and straw mulching with and without herbicide in bare-fallow. *Indian Journal of Ecology* **5**:41-48.

Mohanty M, Sinha NK, Hati KM, Painuli DK and Chaudhary RS 2012. Stability of soil aggregates under different vegetation covers in a vertisol of central India. *Journal of Agricultural Physics* **12** 133-142

Ojeniyi SO and Dexter AR 1984 Effect of soil structure on soil water status. *Soil and Tillage Research* **4** 371- 379

Prihar SS, Singh R, Singh N and Sandhu KS 1979 Yield and water use of dryland crops as affected by mulching in standing maize and fallow. *Experimental Agriculture* **15** 129-134.

10

Water Harvesting for Drought Management in Hot Arid Zone of India

R.K. Goyal

INTRODUCTION

Rainfall is the principal source of water in arid areas, which augments soil moisture, groundwater and surface flows. Agriculture and several of the other economic activities in arid areas depend on rain. This region is devoid of any well-defined perennial river system. Under such circumstances every drop of water becomes very precious. Of the total water use, about 85% of water is used for irrigation and remaining 15% is used for drinking, industrial and other purposes. About 65% of irrigation water and 30-40% of drinking water is subjected to serious losses. Hence, increasing water use efficiency coupled with increasing availability of water through rainwater harvesting and management is of prime importance for sustainable development of arid areas. During the twentieth century, the region has experienced agricultural drought once in three years to every alternate year in one or the other part of the region. The overall probability of drought is 47 per cent. The weather condition, even in average years, for most part of the year remains too dry and inhospitable for successful growth of crops. Rainwater harvesting, its conservation and efficient utilization can solve problem of water scarcity to a greater extend. Rainwater harvesting in small ponds (*Nadis*), underground tanks (*tankas*), *Khadins* (Low lying areas) etc. is an age-old tradition in arid zone of Rajasthan. These traditional rainwater harvesting structures vary in design, shape and size. These traditional methods have undergone several changes for being more economical and efficient. Central Arid Zone Research Institute (CAZRI), Jodhpur since its inception in year 1959 is continuously working for development and refinement of water efficient

---

Principal Scientist (SWCE)
Central Arid Zone Research Institute, Jodhpur 342003 (Rajasthan) India
Email: rkgoyal24@rediffmail.com

techniques for crop production, individuals and for the communities for development in hot arid zone of India.

RAINWATER HARVESTING FOR ARID LANDS

Since primary source of water in arid zone is rainwater, so for any improvement in availability of water, catchment conditions become crucial. Catchment is the base for harvesting rainwater in the form of runoff. It is highly dependent on catchment's shape, size, slope, and type etc. beside rainfall characteristics. Some general modifications can greatly help in enhancing the runoff percentage from catchments.

Techniques for enhancing runoff from catchments

1. Simple earth smoothing and compaction helps increasing runoff from catchment areas. Success is generally greater on loam or clay loam soils. Care must be taken to reduce the slope and/or the length of slope to lessen runoff velocity and thereby reducing runoff.
2. Removal of stones and boulders and unproductive vegetation from catchment helps in uninterrupted flow, enhances runoff to collection site.
3. Land shaping into roads and collection of water in channels.
4. Sandy soils have low water holding capacity. Spreading of clay blanket to the soil surface reduces the infiltration and consequently accelerates runoff.
5. Chemical treatments like wax, asphalt, bitumen and bentonite prevent downward movement of water, which augments runoff.

KHADIN SYSTEM OF RAINWATER HARVESTING FOR CROP PRODUCTION

Khadin is a unique practice of water harvesting, moisture conservation and utilization in hyper arid region of Rajasthan. This system was designed and developed by the Paliwal Brahmins of Jaisalmer (Rajasthan) in the 15th century. It has great similarity with the irrigation methods of the people of Iraq around 4500 BC and later of the Nabateans in the Middle East. A similar system is also reported to have been practiced 4,000 years ago in the Negev desert, and in south western Colorado 500 years ago. The main feature of *khadin* is a very long (100-300 m) earthen embankment built across the lower hill slopes lying below gravelly uplands. Sluices and spillways allow excess water to drain off. The khadin system is based on the principle of harvesting rainwater on farmland and subsequent use of this water-saturated land for crop production. The ratio of farmland and catchment areas is regulated to be about 1:10 so that a suitable moisture supply is uniformly maintained. It is suitable for deep soil surrounded by some natural rock outcrops constituting catchment area. CAZRI has developed *khadin* of 20 ha areas in Baorali-Bambore watershed with surplussing arrangements (Fig. 1). Before construction of *khadin*, uncontrolled runoff from upper catchment used to wash away seeds, fertilizers, and standing crops besides loss of valuable water. After construction of *Khadin*, farmer could take excellent *kharif* and *rabi* crops (Narain and Goyal 2005). Collecting water in a *khadin* aids the continuous recharge of groundwater aquifers. Studies of groundwater recharge through *khadins* in different morphological settings suggest that 11 to 48 per cent of the stored water contributed to groundwater in a single season. This replenishment of aquifers means that subsurface water can be extracted through bore wells dug downstream from the *khadin*. The average water-level rise in wells bored into sandstone and deep alluvium was 0.8 m and 2.2 m, respectively (Khan 1996).

Fig. 1. *Khadin* with waste wier at Baorali-Bambore watershed (Jodhpur)

Design package and guidelines for *Khadin* construction

Central Arid Zone Research Institute, Jodhpur has prepared the design package and guidelines for construction of *khadin* by users' agencies;

1. *Khadin* may be defined as a water harvesting system used for runoff farming on stored soil profile moisture.
2. The catchment may be classified on the basis of infiltration rate. In the areas where infiltration rate is less than 5 cm hr^{-1} may be considered as good catchment. The delineation of catchment should be done on the cadastral/village map or G.T. sheet through reconnaissance survey.
3. The average rainfall of over 30 years available at the nearest rain gauging stations should be considered for working out the catchment yield. Log Pearson III method or strange table should be used.

4. For calculation of flood discharge up to 480 ha area Rational Formula and above 480 ha Dicken's Formula may be used.
5. *Khadin* may be constructed in an area where soil is fine textured, medium to deep with high soil moisture retention capacity. Soil should be free from salinity.
6. In order to have economic design the ponding depth over sill level at the *khadin* bund may vary from 0.65 to 1.10 m with overall average of 0.60 m.
7. The flood lift may be adopted as 0.3 m.
8. During the ponding period from July-October there are wave action therefore, a free board of 0.5 m may be considered.
9. The side slopes of the bunds may be generally kept 2.5:1 (D/S) and 2:1 (U/S). However, these would be governed by the type of soil, angle of repose, bund cross section and its safety factor.
10. The top width of *khadin* bund may be calculated by appropriate formula and not for constructing inspection road.
11. A *murrum* capping of 7.5 cm thick layer be provided over the bund section for protection against wind and rain erosion.
12. The head outlet sluice of appropriate size may be provided in the *khadin* bund for the release of the standing water if any before the *rabi* sowing.

RAINWATER HARVESTING THROUGH *TANKA*

Tankas (small tanks) are underground tanks, found traditionally in most part of western Rajasthan. They are generally built in the main house or in the courtyard. *Tanka* is a circular hole made in the ground, lined with fine polished lime for collection of rainwater primarily from rooftop of individual house. The water collected in small *tankas* is generally used for drinking purposes, however bigger *tanka* can be used for providing supplemental/lifesaving irrigation to horticultural plants. During subnormal rainfall when *tanka* does not get adequate rainwater, water is hauled in camel/bullock cart from nearby wells/*Nadis* to fill the household *tankas*. For rainwater management, CAZRI has designed underground *tanka* of 10 m^3 to 600 m^3 capacities for different rainfall and catchment conditions (Goyal and Issac 2009). These *tankas* were successfully constructed in Jhanwar, Sar, Baorali-Bambore (Jodhpur district) and Kalyanpur (Barmer district) villages. Harvested water of these *tankas* was used to provide lifesaving irrigation to plants. The Benefit cost ratio of *tanka* ranged from 1.25 to 1.40 under different uses (Goyal *et al.* 1995, 2007, Goyal and Sharma 2000).

Tanka should be constructed at appropriate site. If rainwater is to be collected from rooftop, its location should be constructed near to place of intended use. If rainwater is to be collected from natural catchment then *tanka* should be constructed at one side of depression area for maximum runoff and safe disposal of excess water. In arid area of western Rajasthan, a murrum layer is reported in sub surface strata at many places. Special care is needed when *tanka* is to be constructed at these sites. Murrum has a tendency of swelling after getting some moisture and causes cracks especially in sidewalls. To avoid these cracks surrounding of whole *tanka* should have an envelope of 5 cm sand around sidewalls. In case of little leakage from sidewall, sand envelope of 5 cm thickness of sand will absorb the pressure exerted by the swelling of murrum around sidewalls and will prevent the cracks in sidewalls. Circular *tanka* is more economical in comparison to rectangular *tanka* of same capacity in term of cost of materials. Further rectangular *tanka* has tendency for development of cracks in four corners due to uneven distribution of pressure whereas in circular *tanka* pressure distribution is even thus less chance of cracks in sidewalls. Cement concrete is preferred over masonry construction due to cost and life span especially for larger *tanka* of capacity over 1,00,000 litres. However, masonry construction is equally good for small capacity *tanka* and does not require trained workers for construction as in case of cement concrete. Depth of *tanka* should be equal to diameter of *tanka*. Capacity should be based on no. of members in family and livestock. A 21,000 litres capacity *tanka* is sufficient to meet the drinking water requirement of family of 6 persons for round the year (Fig. 2). Approximately 300 m^2 area is needed as catchment under assured minimum rainfall of 180 mm with a runoff coefficient of 0.4 to fill. About 5-10 cm sand filling between masonry wall and earth is recommended for longer life span of *tanka* particularly under clay and heavy soils. To avoid entry of any pollutant from out side, *tanka* is covered at top with stone slabs and cement concrete flooring with an opening for withdrawal of water. Inlet and outlet is provided with weld mesh to avoid entry of any foreign material and small ruminants. A uniform slope of 2-3 per cent towards *tanka* is provided for harvesting maximum possible runoff. The improved design of *tankas* have a lifespan of more than 30 years. Bigger size *tanka*s are more economical. The improved *tanka* design developed at CAZRI has wide acceptability in the region and has been widely replicated in large numbers under Rajeev Gandhi National drinking Water Mission.

Quality of collected rainwater: The cleanest water is always that which falls freely from the sky. The natural water cycle is very efficient in screening out contaminants that are normally found in ground water and other sources. Rainwater does not come in contact with the soil, and so it does not contain contaminants such as harmful bacteria, dissolved salts, minerals or heavy metals. Rainwater is healthy and is soft water so, among other things, less soap is used. Roof-collected rainwater can be made safe and potable by adopting some simple measures such as cleaning of rainwater storage structure and catchment, diversion of first flush and coarse rainwater filters. The quality of rainwater further improves with time after the rain, mainly due to sedimentation and bacteria die-off. It takes an average of 3.5 to 4 days to achieve a 90% reduction in

E.coli numbers. It has been proved that people drinking tank rainwater are at lower risk of many diseases than those drinking public mains water.

Fig. 2. Improved *tanka*

RAINWATER HARVESTING THROUGH *NADI*

Nadis are village ponds used to store runoff water from adjoining natural catchments during the rainy season. In arid Rajasthan *Nadi* system of water harvesting is the oldest practice and still the principal source of water supplies for human and livestock consumption. Across Rajasthan, most *Nadis* have a capacity of 1,200 to 15,000 m^3. Water availability in *Nadi* ranged from 2-12 months after the rains. Since *Nadis* received runoff from sandy and eroded rocky basins, large amounts of sediments used to deposit regularly in them, resulting in quick siltation. High evaporation and seepage losses through porous sides and bottom, heavy sedimentation due to biotic interference in the catchment and contamination are major bottlenecks. Evaporation losses ranged from 55 to 80 per cent of the total losses in various environments. Seepage losses are greatest during the rainy season (July-September) when *Nadi* is completely filled. To overcome these problems, CAZRI has developed design for improved *Nadis* with LDPE lining on sides and bottom keeping surface to volume ratio 0.28 and provision of silt trap at inlet.

The site selection of *Nadi* is based on availability of natural catchment and its runoff potential (Fig. 3). The location of the *Nadi* had a strong bearing on its storage capacity due to catchment and runoff characteristics. *Nadis* are 1.5 to 4.0 m deep in dune areas and those in sandy plains vary from 3 to 12 m. In addition, planting suitable tree species around the *Nadi* creates an oasis in the desert and improves the local environment.

Fig. 3. Improved *Nadi* with inlet and outlet

Fig. 4. Circular micro-catchment for tree establishment

RAINWATER HARVESTING THOROUGH MICRO-CATCHMENT FOR TREE ESTABLISHMENT

Micro-catchment technique is particularly suitable for establishment of trees. In this technique a circular catchment of 1 to 1.5 m radius is constructed around the tree (Fig. 4). The catchment is compacted by roller or any other heavy machine. A slope of 5-10 per cent is provided in catchment towards tree for directing flow of water. The catchment can also be lined with locally available materials such as polythene sheets, lime mortar, stone pieces, grasses, etc. for higher runoff generation. It is reported that the plants with micro-catchment have better chances of establishment in rainfed conditions as compared to conventional plantation technique (Ojasvi et al 1999). In another study Sharma et al (1986) suggested that conversion of canopy area into runoff catchment may be just sufficient for improving its soil moisture profile.

CONTOUR VEGETATIVE BARRIERS

Contour vegetative barriers are hedgerows of perennial grasses or shrubs planted at a regular interval on contours for conserving soil and water in sloping lands. Suitable grass species are grown along contours at suitable vertical interval to intercept part of runoff and to control erosion in agricultural fields having flat to slight undulating topography. The contour vegetative barrier moderates the velocity of overland flow and traps silt at low cost, and augment production of food, fuel and fodder or fibre from lands by growing suitable vegetation species. In recent years, contour vegetative barriers have found acceptability among the farmers as these are cheaper over mechanical measures and are protective while being productive. Contour vegetative barriers can be easily established across a wide spectrum of soil-climatic conditions. Selection of species depends upon purpose of barrier, site-specific conditions, particularly soil and climatic variables. The spacing between plant-to-plant and row-to-row is governed by vegetation species to be planted as barrier. In general the plant-to-plant spacing is kept at 20 to 30 cm. Predetermined or 0.5 to 1 m vertical interval between the barriers has been found effective for soil and water conservation. Generally, paired row of barriers planted in staggered form across the slope proves more effective.

Generally, dominant grass or shrub species of the region should be preferred for vegetative barrier. Among grasses *Cenchrus ciliaris, Cenchrus setigerus, Saccharum bengalense, Vetiveria zizanioides, Lasiurus sindicus, Panicum antidotale* and *Panicum turgidum* can be effectively used for soil and water conservation in arid areas. Shrubs like *Leptadenia pyrotechnica, Ipomoea carnea* and *Euphorbia antisyphylitica* can also provide good protection against water and wind erosion. Contour vegetative barriers of *C. jwarancusa, C. ciliaris* and *C. setigerus* transplanted at 0.30 m apart on contours at 0.6 to 1.0 m vertical interval in sandy loam soil of Jodhpur (Rajasthan) have performed well and formed effective barriers in reducing soil erosion and increasing soil moisture storage. In a study conducted during 1992-1994 at 19 farmers' fields near Jodhpur rooted slips of local eight species of perennial grasses (*C. ciliaris, C. setigerus, C. jwarancusa, L. sindicus, P. antidotale, P. turgidum, S. bengalense* and *V. zizanioides*) and seedling of six species of shrubs (*Agave americana, Aloe barbadensis, Barleria prionitis, E. antisyphylitica, I. carnea* and *L. pyrotechnica*) were transplanted at 1 m vertical interval on contours across the slope. Result indicated that perennial grass species performed the best and formed effective barrier against soil erosion. Runoff volume and specific peak discharge were reduced by 28 to 97 per cent and 22 to 96 per cent, respectively (Sharma et al

1999, Tiwari and Kurothe 2006). In another study conducted at Kalyanpur (Barmer district) during 1998, vegetative barrier of *L. sindicus, Saccharum munja* and *Cassia angustifolia* were established at horizontal interval of 30 m. The moisture data revealed 36.5, 72 and 54.2 per cent higher moisture storage as compared to control in *C. angustifolia, L. sindicus* and *S. munja*, respectively (Gupta and Rathore 2002).

REFERENCES

Goyal RK and Sharma AK 2000 Farm Pond: A well of wealth for the dryland dwellers. *Intensive Agriculture*, **XXXVIII(5-6)** 12-14 & 26

Goyal RK and Issac VC 2009 *Rainwater Harvesting through Tanka in Hot Arid Zone of India*. Central Arid Zone Research Institute, Jodhpur, CAZRI Bulletin, 33 pages

Goyal RK, Ojasvi PR and Bhati TK 1995 Economic evaluation of water harvesting pond under arid conditions. *Indian Journal of Soil Conservation* **23(1)** 74-76

Goyal RK, Bhati, TK and Ojasvi PR 2007 Performance evaluation of some soil and water conservation measures in hot arid zone of India. *Indian Journal of Soil Conservation* **35(1)** 58-63

Gupta JP and Rathore SS 2002 *Biomass Production and Rehabilitation of Degraded Lands in Arid Zone,* Central Arid Zone Research Institute, Jodhpur,20 pages

Khan MA 1996 Inducement of groundwater recharge for sustainable development. *In Proceedings of the 28th Annual Convention*, Indian Water Works Association, Jodhpur, India, 147-150 pages

Narain P and Goyal RK 2005 Lead paper on 'Rainwater harvesting for increasing productivity in arid zones'. National Symposium on "Efficient water management for eco-friendly sustainable and profitable agriculture". *Symposium Abstract.* Organized by Indian society of water management and Indian Agriculture Research Institute from Dec. 1-3, 2005 at Water Technology Centre, New Delhi. 141-142 pages

Ojasvi PR, Goyal RK and Gupta JP 1999 The micro-catchment water harvesting techniques for the plantation of Jujube (*Zizyphus mauritiana*) in an agroforestry system under arid conditions. *Agricultural Water Management* **41** 139-147

Sharma KD, Joshi NL, Singh HP, Bohra DN, Kalla AK and Joshi PK 1999 Study on the performance of contour vegetative barriers in an arid region using numerical models. *Agricultural Water Management* **41** 41-56.

Sharma KD, Pareek OP and Singh HP 1986 Micro-catchment water harvesting for raising jujube orchards in arid climate. *Transactions of American Society of Agriculture Engineering* **29(1)** 112-118

Tiwari SP and Kurothe RS 2006 Effect of vegetative barriers on soil and nutrient losses at 2% slope on agricultural lands of reclaimed Mahi ravines. *Indian Journal of Soil Conservation* **34(1)** 37-41.

11

In situ Moisture Conservation for Mitigating Drought

R.K. Singh

INTRODUCTION

Population growth and the improvement of living standards have increased water demand worldwide. This situation is to a great extent attributable to agriculture, which is the world's largest water consumer (OECD 2013, Ward and Pulido-Velazquez 2008). Drought occurs widely and frequently in India and around 68% area is prone to it. Of this, 35% receiving rainfall between 750 mm and 1125 mm is considered drought prone while 33% receiving less than 750 mm is chronically drought prone. The major water-related challenge for rainfed agriculture in arid, semiarid and dry sub humid regions is the extreme variability in rainfall, characterized by few rainfall events, high-intensity storms and high frequency of dry spells and droughts. In the arid regions it is the absolute amount of water that constitutes the major limiting factor in agriculture. In the semiarid and dry sub humid tropical regions quantum of seasonal rainfall is generally adequate to significantly improve yields but managing extreme rainfall variability over time and space is the largest water challenge. Only in the dry semiarid and arid zones, even when considering the standard deviation around the mean, is absolute water stress common. In general, only 70%–80% of the rainfall is available to plants as soil moisture, and on poorly managed land the share of plant-available water can be as low as 40%–50% (Falkenmark and Rockstrom 2004). This leads to agricultural dry spells and droughts, which are due primarily to management-related problems with the on-farm water balance and are thus an indicator of large opportunities to improve yields through better water management. Agarwal (2000) opined that India would not have to suffer from droughts if local water balances were better managed. Even during drought years better rainfall

---

Senior Scientist
Central Arid Zone Research Institute, Jodhpur 342003 (Rajasthan) India
Email: ranjay_2005@rediffmail.com

management has benefited Indian farmers; villages benefiting from watershed management projects increased food production and market value by 63% compared with those without such projects (Wani et al 2006).

Water harvesting is applied in arid and semiarid regions where rainfall is either not sufficient to sustain good crop and pasture growth or where, due to the erratic nature of precipitation, the risk of crop failure is very high. Water harvesting can significantly increase plant production in drought prone areas by concentrating the rainfall/runoff in parts of the total area. The intermittent character of rainfall and runoff and the ephemerality of floodwater flow require some kind of storage. There might be some kind of interim storage in tanks, cisterns or reservoirs or soil itself serves as a reservoir for a certain period of time. Glantz (1994) has pointed out that agricultural droughts, caused primarily by a poorly performing water balance, are more common than meteorological droughts. Evidence from water balance analyses on farmers' fields around the world shows that only a small fraction of rainfall (generally less than 30%) is used as productive green water flow (plant transpiration) supporting plant growth (Rockstrom 2003).

Surface water is naturally replenished by precipitation and naturally lost through discharge to evaporation and sub-surface seepage into the groundwater. Surface water storage includes water accumulated on the soil surface or underground, intercepted water, *i.e.,* the rainfall quantity retained by vegetal coverage, and water retained in depressions (flooded plains, lakes, swamps, pools). For surface water conservation *in situ*, the art and science of "collecting water where it falls" needs to be revived to meet modern freshwater needs adequately, equitably and sustainably. Water harvesting technology is especially relevant to the semi arid and arid areas where the problems of environmental degradation, drought and population pressures are most evident and given the increasing frequency and severity of droughts and population growth, it becomes increasingly important to systematically explore the unrealized potential of anti-drought measures including water harvesting, water conservation and efficient water utilization. While irrigation may be the most obvious response to drought, it has proved costly and can only benefit a fortunate few.

Water harvesting may be *in situ,* where runoff is concentrated into catchments surrounding a tree or group of trees, or *ex situ* where runoff from large rocky areas is diverted directly to nearby orchards or after collection in reservoirs or farm ponds.

For *in situ* water conservation practices that reduce runoff by controlling water movement over the surface need to be adopted. The principle of these practices is to minimize the concentration of runoff volume and to slow down the runoff velocity, allowing the water more time to soak into the soil, limiting its capacity to transport soil particles and diminishing its ability to cause erosion. The land use can be single or in combination of agriculture, horticulture, forestry and pasture. The conservation measures may include contour farming, contour bunding, contour trenching, terrace farming, vegetative barriers, strip cropping, etc. with their suitability depending on slope gradient, soil type and depth and rainfall. Other *in situ* water conservation techniques include no-tillage and deep tillage farming, compartmental bunding, ridges and furrow system, micro catchment techniques, tie ridging methods etc. Generally, two kinds of practices are followed in conserving and managing water *in situ* in combination with land management practices.

1. Reduce runoff through improved infiltration capacity and soil transmission characteristics.
2. Reduce runoff by controlling water movement over the surface.

Different kinds of techniques falling under these categories for *in situ* water conservation are explained in this paper

VEGETATIVE MULCHING

It is a system of maintaining a protective cover of vegetative residues such as straw and stubbles on the soil surface at all times. The system is particularly valuable where a satisfactory plant cover cannot be established at the time of year when erosion risk is at its maximum. The beneficial effects of mulching include protection of the soil surface against raindrop impact, decrease in flow velocity by imparting roughness, and improved infiltration capacity. The quantity of mulch required for maintenance of favourable infiltration capacity and structural stability depends on the rate of residue decomposition, climate, soil properties, relief and rainfall characteristics. Favourable effects of mulching on soil and water conservation have been reported for the arid and semi-arid regions (Lawes 1966, Djokoto and Stephens 1961, Adeoye 1985). Adequate or durable mulching materials are less readily available in these regions with their long dry season. There are however, other techniques providing utilizable mulch by using cover crops, alley cropping and no tillage.

COVER CROPS

Cover crops can improve soil quality by increasing soil organic matter levels through the input of cover crop biomass over time. Increased soil organic matter enhances soil structure as well as the water and nutrient holding and buffering capacity of soil (Patrick et al 1957). With increased water infiltration, the potential for soil water storage and the recharging of aquifers can be improved (Joyce et al 2002). It is the most satisfactory method of building up the organic content of soil. The effectiveness of cover crops in soil and water conservation, however, depends on species characteristics including ease of establishment, vigour depth of rooting, rapidity of establishment of surface cover etc. Singh and Nath (1997) reported that under semi arid conditions of Pali (Rajasthan) 4.36 q/ha mung bean (*Vigna radiata*), 10.23 q/ha cluster bean and 1.88 q/ha sesame can be produced under *ber* (cv. Seb) based farming system during *kharif.* The yields of both *ber* and the intercrops were higher than in monoculture and no adverse effects were observed up to five years after planting the *ber* trees (Singh 1997 Patel et al 2003). Cover cropping with *kulthi* (*Dolichos biflorus*) was found to increase water holding capacity of light soils as a result of increased organic carbon content in arid regions of the tropics (Pareek and Nath 1996).

DEEP TILLAGE

Under semi-arid conditions in India conventional tillage was found superior to no tillage, reduced tillage or mulching with a number of crops i.e. sun hemp (*Crotalaria juncea*), barley (*Hordeum vulgare*), mustard (*Brassica juncea*) and chickpea (*Cicer arietinum*) grown in the dry season (Rao et al 1986). In West African semi-arid regions, soil inversion and deep ploughing was found superior to no tillage in increasing plant-available water and crop yields (Nicou 1977; Charreau 1972, 1977). Higher response of tillage over no tillage or greatly reduced tillage was found over a variety of soils (Karaca et al 1988, Prihar and Jalota 1988, Willcocks 1988). In a study conducted at Bikaner, Rajasthan, the moth bean–cluster bean cropping system recorded 20% higher equivalent yields, than conventional tillage (Rathore et al 2014). Another study conducted at Pali-Marwar on taramira yield and water productivity, deep tillage during monsoon increased the seed yield of taramira by 7.6 and 3.4% over pre monsoon deep tillage in the year 1999-00 and 2001-02 (Regar et al 2009).

STRIP CROPPING

This is farming of sloping land in alternate contoured strips of intertilled row crops and close-growing crops (e.g. a cover crop or grass) aligned at right angles to the direction of natural flow of runoff. The close growing strip slows down runoff and filters out soil washed from the land in the intertilled crop. Usually, the close growing and intertilled crops are planted in rotation. Strip cropping provides effective erosion control against runoff on well drained erodible soils on 6 to 15% slopes. The width of the strips is varied with the erodibility of the soil, and slope steepness. In effect, it employs several good farming practices such as crop rotation, contour cultivation, stubble mulching, etc. In a study conducted at Faa village of Jordan, characterized by arid conditions, the barley production as biomass was highest using strip cropping ratios of 1:1, 1:2, and 1:3 as opposed to a conventional farm system (Abu- Nukta et al 2009).

VEGETATIVE BARRIERS

Vegetative materials (e.g. vetiver) planted across the slope arrest the runoff. This is suitable for various slopes of land. The use of vegetative barrier can be a permanent viable alternative to costly engineering measure for soil and water conservation. A judicious choice should be made between adapted species such as *Napier*, which are more efficient at capturing rainwater but compete with crops for available water, and species such as *Leucaena* which seem to compete less for water with the crops, but are less efficient at capturing rainwater (Serraj and Siddique 2012) .

CONTOUR FARMING

Contour farming involves aligning plant rows and tillage lines at right angles to normal flow of runoff. It creates detention storage in the soil surface horizon and slows down the runoff thus giving the water time to infiltrate into the soil. The effectiveness of contour farming in water and soil conservation depends not only on the design of the system but also on soil, climate, slope aspect and land use. Contour bunds are another important physical measure. These are earth banks, 1.5 to 2 m wide, forming buffer strips at 10 to 20 m intervals across the slope.

RIDGES AND FURROW SYSTEM

This is laid out in the soils having approximately one per cent slope. Furrows serve as effective means of surface drainage. This system is highly effective in medium to high rainfall areas (700-1200 mm). The ridge-furrow system is a commonly used physical conservation practice. Ridge-tying is also an effective soil and water conservation system especially in arid and semi-arid regions. Mounds and tied mounds are also effective in conserving water and soil. Tie ridges or mounds to permit more rainwater to infiltrate is an effective system in drier areas (< 1000 mm annual rainfall) on gentle slopes (< 7%) but not in wet years or more humid areas.

BROAD BED AND FURROW SYSTEM

It essentially consists of land smoothing with less soil disturbances after primary and secondary tillage and formation of graded wide beds separated by furrows connected to grassed water ways. The beds and furrows are laid out on gradient of 0.4-0.8 per cent for optimum performance. The beds are 120 cm wide and 15 cm deep. This system is suitable for lands with a slope not exceeding 3 per cent. This system permits safe disposal of runoff and tends to conserve soil and water *in situ*.

TERRACE FARMING

Terrace farming involves earthworks made at right angles to the steepest slope consisting of an excavated channel on the uphill side, the spoil from which forms a bank on the downhill side. It converts a steep slope into a series of steps with horizontal or nearly horizontal ledges and vertical walls of stone, brick or timber between ledges.

MICRO/CIRCULAR CATCHMENT TECHNIQUE

Micro-catchment is one of the major forms of direct water conservation systems. It is a method of collecting surface runoff from a small catchment area and storing it in the root zone of an adjacent infiltration area/basin. In micro-catchment based cropping, rainwater is concentrated in a small portion of the cultivated area. The tree crops are deep rooted and can utilize the moisture stored in the sub-stratum and hence form a better option for micro-catchment based farming in sandy soil situations (Sharma et al 1986).

The circular catchment technique is one of the most feasible methods for *ber* under rainfed condition, which helps in collecting more rainwater and makes available for successful establishment, growth and yield of the transplanted plants (Gupta 1984). The technique involves the construction of compacted circular catchment of 1.5 m radius with 5-10% slope around the transplanted plants. The results of the field trial showed that even during low rainfall years from 1986-88, circular catchment technique increased the mean soil profile moisture storage. Sharma et al (1982) suggested that highest *ber* yield was obtained when 0.5% and 5% slopes had 8.5 m and 7.0 m length of run and 72 m^2 and 54 m^2 area per tree, respectively.

COMPARTMENTAL BUNDING

The principle behind the compartmental bunding is to increase the infiltration by reducing the rate of runoff, temporarily impounding the water on the surface of the soil to increase the opportunity time for infiltration and modifying the land configuration for inter plot water harvesting.

A field experiment was conducted during 1998-99 and 1999-2000 in Karnataka, India to study the response of desi cotton (*Gossypium herbaceum*) to *in situ* moisture conservation and organic manures. Under the rainfed condition higher seed cotton yield and net returns could be achieved by adopting *in situ* moisture conservation practices such as compartmental bunding, tied ridges and furrows and broad bed and furrow alongwith 5 tonnes/ ha farmyard manure or poultry manure without impairing the soil fertility (Hulihalli and Patil 2006).

CONCLUSION

In situ surface water management aims to enhance surface detention storage, thus allowing the water more time to soak into the soil, and to conduct away unavoidable runoff non-erosively. A wide range of *in situ* water management techniques are in use. Because the choice of appropriate measures depends on soil, topography, climatic and socio-economic considerations, generalization about their applicability is difficult. The most appropriate strategy for surface water management depends upon identifying the key factors influencing runoff and applying appropriate techniques based on acquired knowledge. Since *in situ* water management strategies are often relatively cheap and can be applied on any piece of land, they should be optimized before water from external sources is considered.

REFERENCES

Abu-Nukta, Ali Sertyesilisik, Begum and Alkhaddar Rafid 2009 Cropping systems as water harvesting techniques for barley production in arid and semi-arid areas in Jordan, *Journal of Facilities Management* **7(4)** 319-331

Adeoye KB 1985 Effect of some management practices on soil crusting in cultivated savanna soils. In: *Soil Fertility, Soil Tilth and Post-clearing Land Degradation in the Humid Tropics* (Eds. R.A. Sobulo and E.J. Udo), SSSN Proc. 302-310 pages

Agarwal A 2000 Drought? Try Capturing the Rain. Briefing Paper. New Delhi: *Centre for Science and Environment*

Charreau C 1972 Problèmes posés par l'utilisation agricole des sols tropicaux par les cultures annuelles, *Agronomie Tropicale* **27** 901-929

Charreau C 1977 Controversial points in dryland farming practices in semi-arid West Africa. *In Proceedings of Symposium on Rainfed Agricultural Semi-arid Regions,* (ed) GH Cannell, University of California, Riverside 313-360 pages

Djokoto RK and Stephens D 1961 Thirty long-term experiments under continuous cropping in Ghana. I. Crop yields and responses to fertilizers and manures. *Empire Journal of Experimental Agriculture* **29** 181-195

Falkenmark M and Rockstrom J 2004 Balancing Water for Humans and Nature: The New Approach in Ecohydrology, London: Earthscan.

Glantz MH 1994 *Drought Follows the Plough: Cultivating Marginal Areas*. Cambridge, UK: Cambridge University Press.

Gupta JP 1984 Moisture conservation in arid lands, *Indian Farming* **34 (7)** 5-6

Hulihalli UK and Patil VC 2006 Effect of *in situ* moisture conservation practices and organic manures on yield and quality of cotton *(Gossypium herbaceum* L.) grown on vertisol, *PKV Research Journal* **30** 135-140

Joyce BA, Wallender WW, Mitchell JP, Huyck LM, Temple SR, Brostrom PN and Hsiao TC 2002 Infiltration and soil water storage under winter cover cropping in California's Sacramento Valley, *Transactions of the American Society of Agricultural Engineers (ASAE)* **45** 315-326

Karaca M, Guler M, Durutan N, Pala M and Unver I 1988 Effect of fallow tillage systems on wheat yields in Central Anatolia, Turkey. In: *Challenges in Dryland Agriculture*. (eds) P.W. Unger et al., Proc. Int. Confer. on Dryland Agriculture, Amarillo/Bushland, Texas. 131-133 pages

Lawes DA 1966 Rainfall conservation and the yields of sorghum and groundnuts in northern Nigeria, *Experimental Agriculture* **2** 139-146

Nicou R 1977 *Le travail du sol dans les terres exondées du Sénégal: motivations, contraintes*. Centre National de Recherche Agronomique (CNRA), Bambey, Senegal. Mimeo. 50 page

OECD 2013 Water Security for Better Lives (Report), OECD Studies on Water. Organization for Economic Co-operation and Development, Paris, France.

Pareek O P and Nath V 1996 Ber. In: *Coordinated fruit research in Indian arid zone - A two decades profile* (1976-1995), Central Arid Zone Research Institute, Jodhpur

Parr JF, Papendick RI, Hornick SB and Meyer RE 1990 The use of cover crops, mulches and tillage for soil water conservation and weed control. In: *Organic-matter*

Management and Tillage in Humid and Sub-humid Africa IBSRAM Proceedings No.10. Bangkok: IBSRAM 246-261 pages

Patel BM, Patel SI, Patel SK and Patel SB 2003 Intercropping studies in ber *(Ziziphus mauritiana* Lamk.), *Agricultural Science Digest* **23(2)** 113-115

Patrick WH, Haddon CB and Hendrix JA 1957 The effects of longtime use of winter cover crops on certain physical properties of commerce loam, *Soil Science Society of America* **21** 366-368.

Prihar SS and Jalota SK 1988 Role of shallow tillage in soil water management. In: *Challenges in Dryland Agriculture*. (Eds. P.W. Unger et al.), Proc. Int. Confer. on Dryland Agriculture, Amarillo/Bushland, Texas 128-130 pages

Rathore VS, Singh JP, Meel B and Nathawat NS 2014 Agronomic and economic performances of different cropping systems in a hot, arid environment: A case study from North-western Rajasthan, India, *Journal of Arid Environments* **105** 75-90

Regar PL, Rao SS and Joshi NL 2009 Effect of in-situ moisture conservation practices on productivity of rainfed taramira *(Eruca sativa)* in arid Rajasthan, *Indian Journal of Soil Conservation* **37 (3)** 197-200

Rockstrom J 2003 Water for Food and Nature in Drought-prone Tropics: Vapour Shift in Rain-fed Agriculture, *Royal Society Transactions B Biological Sciences* **358 (1440)** 1997–2009.

Roose EJ 1988 Soil and water conservation lessons from steep-slope farming in French-speaking countries of Africa. In: *Conservation Farming on Steeplands* (Eds. W.C. Moldenhauer and N.W. Hudson), SWCS, Ankeny, IA. 129-139 pages

Serraj Rachid and Siddique Kadambot HM 2012 Conservation agriculture in dry areas, *Field Crops Research* **132**: 1–6

Sharma KD, Pareek OP and Singh HP 1982 Effect of runoff concentration on growth and yield of jujube, *Agricultural Water Management* **5 (1)** 73-84

Sharma KD, Pareek OP and Singh HP 1986 Micro-catchment water harvesting for raising *jujube* orchards in an arid climate, *Transactions of the American Society of Agricultural Engineers (ASAE)* **29 (1)** 112-118

Singh RS 1997 Note on the effect of intercropping on growth and yield of ber *(Ziziphus mauritiana* Lamk.) in semi-arid regions, *Current Agriculture* **21(1/2)** 117-118

Singh RS and Nath V 1997 Intercropping in ber, *Intensive Agriculture* **34(11 & 12)** 14-17

Unger PW, Langdale GW and Papendick RI 1988 Role of crop residues - improving water conservation and use. In: *Cropping Strategies for Efficient Use of Water and Nitrogen* (Eds. W.L. Hargrove. *American Society of Agronomy Special Publication* **51** 69-100

Wani SP, Ramakrishna YS, Sreedevi TK, Long TD, Wangkahart T, Shiferaw B, Pathak P, and Kesava Rao AVR 2006 "Issues, Concepts, Approaches and Practices in Integrated Watershed Management: Experience and Lessons from Asia." In: *Integrated Management of Watersheds for Agricultural Diversification and Sustainable Livelihoods in Eastern and Central Africa: Lessons and Experiences from Semi-Arid South Asia* (Eds. B. Shiferaw and K.P.C. Rao) Proceedings of the International Workshop held at International Crops Research Institute for the Semi-Arid Tropics, Nairobi, 6–7 December 2004. Andhra Pradesh, India: International Crops Research Institute for the Semi-Arid Tropics.

Ward FA and Pulido-Velazquez M 2008 Water conservation in irrigation can increase water use, *Proceedings of the National Academy of Sciences* **105**: 18215–18220

Willcocks TJ 1988 Tillage systems for sustainable rainfed crop production in semi-arid regions of developing countries. *In Challenges in Dryland Agriculture* (eds) PW Unger et al Proceedings of International Conference on Dryland Agriculture, Amarillo/Bushland, Texas. 131-133 pages.

12

Strategies for Limited Water Use in Field Crops

Anurag Saxena

INTRODUCTION

Efficient use of water in agriculture is an important issue throughout the world. Agriculture, the largest water user, is facing a challenge to produce more food with minimum water. This requires improvement in water productivity (WP) and water use efficiency (WUE). This is more pertinent to tropical country like India that experiences the variation in climate and rainfall. India, though has achieved self-sufficiency in food with the help of efficient management of her natural resources leading to green revolution but arid and semi-arid regions are still lagging behind. It is necessary to optimise the use of water and at the same time increase the water productivity (Saxena et al 2014). This can be achieved by optimizing irrigation, improving water delivery, application methods and rain water conservation. Water can be conserved at a watershed or regional or local level for other uses only if evaporation, transpiration, or both are reduced. Precision irrigation technologies for sprinklers and micro-irrigation system enables farmers to apply water and agrochemicals more precisely. The most effective means to conserve water is deficit irrigation strategies that are supported by advanced irrigation system.

Harsh climatic conditions in arid regions lead the farmers to grow only one crop during rainy season that to depending on the rainfall. The average annual rainfall of western arid region of India is 317 mm. The rainfall is highly variable and erratic with frequent drought spells. The number of rainy days vary from 10 to 25. Groundwater is very deep, saline at many places and expensive to use. Injudicious water use on undulating highly permeable sandy soils through conventional irrigation resulted in fall of groundwater by 0.6 – 1.0 m annually. The situation of overexploitation of groundwater is more serious in the region where out of 11 districts, 6 are in category of over exploited and remaining 5 are

Principal Scientist, Central Arid Zone Research Institute, Jodhpur - 342 003
Email: anurag.saxena@icar.gov.in; drasaxena@gmail.com

in category of semi critical zone. On the other hand indiscriminate use of scarce water through conventional irrigation management practices led to exhaustion of ground water resources and development of water logging in canal command area. Hence, conservation and efficient management of limited water is the need of the hour for achieving sustainable production for longer period on light textured soils of arid and semi-arid regions.

The first option concerns to the need for improving crop yield; the second one intends to increase the transpiration of water supply against the evaporation; the third aims to utilize efficiently the water resources (reservoirs, streams or groundwater sources). All these lead to the improvement in management practices which use less water for irrigation, decrease evaporation losses, optimize fertilizer supply, pest control, energy consumption, soil conditions, etc. This has more importance in arid and semi-arid regions with limited water supply, where the farmers are frequently constrained to apply deficit irrigation strategies and manage water supply in accordance with the sensitivity of crop's growing stages to water stress. There are several approaches for improving the crop productivity of water including replacing high water consuming crops with lower-consuming ones and adopting management and systems improvements to increase productivity per unit of water consumed. Efficiencies are increased when the total amount of water consumed by crops, evaporation and other users can be reduced. The available water resource within a basin or sub basin can also be effectively conserved for other uses by improving efficiencies to reduce the unusable water losses.

WATER FOOTPRINT

The water footprint of a product is the total volume of freshwater used to produce the product, summed over the various steps of the production chain. The water footprint of a product refers not only to the total volume of water used; it also refers to where and when the water is used.

Green water footprint: Volume of rainwater consumed during the production process. This is particularly relevant for agricultural and forestry products where it refers to the total rainwater evapotranspiration.

Blue water footprint: Volume of surface and groundwater consumed as a result of the production of a good. Consumption refers to the volume of freshwater used and then evaporated or incorporated into a product.

Grey water footprint: The grey water footprint of a product is an indicator of freshwater pollution that can be associated with the production of a product over its full supply chain.

The intrinsic characteristic of water dynamics in soil cannot be altered to a large degree but availability can be modified enormously by water conservation and water management practices. During the complete life cycle of a plant the most dynamic property is water content of soil. Since water that is conserved or supplemented becomes available to the plant through the soil, monitoring of water content under field condition is of paramount importance. The whole life cycle of plant revolves around extraction of water from the soil through its root network and transpiring through the leaves in response to evaporation demand set up by the atmosphere. Any shortage of water in the soil is ultimately reflected by the growth of the plants in some form or another. Increasing water supply available to plants or enhancing water use relative to other losses and making efficient use of limited water are the three major concepts in optimizing water use in agriculture.

RAIN WATER MANAGEMENT

In arid and semi-arid parts, experiencing rainfall from 150-350 mm and 350-550 mm respectively, crop production is adversely affected due to low and erratic distribution. In last two decades researcher and planners have given lot of attention to dry land farming in such areas. Various regions specific agro-techniques have been evolved.

Choice of crops based on rainfall and moisture availability period

Improving on-farm irrigation efficiencies will usually not save water and in fact increase total water use because both higher irrigation uniformities and increased inputs improve total yields. Farmers shift to higher value, higher water-using crops because the improved irrigation system renders it more feasible. In dryland areas crop production depends on amount and distribution of rainfall. The options may be to include drought-tolerant cultivars, by increasing farmer's ability to optimize irrigation amounts in time and space utilizing site-specific irrigation techniques and by including deficit irrigation. In arid and semi-arid areas where frequent partial season droughts occur, crops that mature more quickly such as small grains, various pulse crops, deep rooted, drought-resistant crops would be grown to maximize use of precipitation stored in the soil. Most of the farmers still grow long duration traditional varieties under conventional practices. The productivity of these crop varieties is affected due to uncertain rainfall. Hence, there is a need to develop short duration varieties and agronomic practices matching to rainfall pattern. On the basis of climatic data all over India, suitable crops and varieties for early, normal and late onset of monsoon, improved inter cropping, time of sowing and agronomic practices have been identified and developed. In arid part of Rajasthan short duration varieties of pearl millet, clusterbean, moth bean, mung bean and sesame have been developed to minimize the risk of crop failure. With the early onset of monsoon, pearl millet and sesame get preference. While with the late onset of monsoon clusterbean, mung bean, and moth bean get preference. Since the moisture evaporates very fast, it has been suggested to complete the sowing immediately after rains. Certain simple agronomical practices like optimum tillage, administration of organic manure, suitable cropping pattern, and strip cropping are effective in retaining soil fertility as well as giving satisfactory crop yield.

Increasing water use relative to other losses

Evaporation of water directly from soil surface is an important component of the total water use of crops. Evaporation constitutes 19-46% of the total water on vertisols, 21% on alfisols and 12-18% on loamy sand soils. Reducing the evaporation mulching is an effective tool. Soil evaporation can be substantially reduced through minimum tillage (Saxena et al 1997) and mulching using materials or crop residues. Mulching of open land surface is achieved by spreading stubble, trash or any other vegetation. The objectives of mulching are to minimize splash influence of rain drops on base surface; reduce evaporation; increase absorption of the rainfall; obstruct surface flow thereby retarding erosion and allow microbiological changes to occur at optimum temperature. Soil evaporation can also be reduced by achieving a rapid ground cover through early crop vigour and fertilizers have been shown to help. Top surface of loamy sand soils acts as self-mulch after drying. In such conditions, mulch may not be useful practice in arid region, however, many a times droughts occur early in the vegetative stage. In such situation use of mulch immediately after sowing delay the process of top soil surface drying which helps in promoting better plant establishment and results in higher yield. Polyethylene mulch was highly effective in controlling evaporation losses (Table 1, Gupta 1978, 1980). Daulay et al (1979) also reported beneficial effects of mulch in pearl millet. Application of grass mulch (6 t ha^{-1}) brought 200 per cent increase in the yield of green gram, dew gram and clusterbean (Gupta and Gupta, 1983). The best opportunity for water saving lies in methods increasing transpiration relative to other losses from the soil surface due to canopy shading. The elimination of weeds, optimum plant population and spacing and application of fertilizer/manure helped in better canopy growth and higher WUE of crops. Generally close row spacing and higher plant population is recommended in high rainfall areas or irrigated condition, but in arid region wide row spacing and low population is recommended. Paired row planting system helps in better canopy growth and hence evaporation from soil surface is restricted.

Table 1. Mulching improves Yield and WUE of Pearl millet

Mulch	Water Use (ET) mm	Yield (kg ha^{-1})	WUE (kg ha^{-1} mm^{-1})
Polyethylene	279	2900	10.4
Bajra husk	269	2300	8.5
No mulch	291	1740	6.0

Tillage makes soil loose and hence prone to erosion. Adoption of minimum tillage practices also help with early and rapid sowing. As a result, better matching of crops and environment, coupled with the responsive cultivars to increased inputs, disease control and other agronomic changes, yields improvement can be achieved. Timing and depth of tillage are the two important factors, which need special attention (Saxena et al 1997). Tillage should be done immediately before the crop season to take advantage of one or two early showers for land preparation. In arid region, land tilled into ridges and furrows across the wind direction has been found to reduce the effects of wind erosion during the summer months. Farm lands should be levelled for efficient management of water. Level lands allow more infiltration, thus increasing soil moisture and leaching. This in turn reduces run-off and hence soil erosion. Levelling of irregular land is done by the cut and fill method. Soil from the elevated portions is removed and placed in the depressed portions to obtain a level land. In strip cropping two or more crops are grown in alternate strips preferably across the slope. One strip of erosion permitting crop should be alternated with another strip of relatively more erosion-resisting crop. In strip cropping, the entire land remains in a kind of balance. Soil eroded from one strip is retained by the next strip and the overall fertility of the land is maintained.

Management of limited irrigation water

In arid region water is limited and land is vast, hence water management aims to maximize production per unit of water rather per unit of land. Some of the technologies like modest, extensive, deficit and pressurized irrigation have been developed to maximize the production in arid region.

Rules for Extensive and Deficit irrigation

- Practice extensive/deficit irrigation only on relatively deep soils that have moderately high water holding capacity
- Increase the contribution of precipitation to crop water needs
- Consider modifying cultural practices
- Flexible planting dates
- Deep tillage for deeper rooting
- Selection of irrigation system (Drip, Sprinkler)

Extensive irrigation: Extensive irrigation approach seeks to apply a small quantity of water over a large area rather large quantity of water over a small area. Production per unit land may decrease but production per unit water may increase. Water requirement of wheat and mustard is 840 and 250 mm per hectare to produce maximum yield. When the same water applied optimally in 3, 1.5 and 4.0 ha land in wheat and mustard, respectively gave less production per unit of land but total productivity per unit of water was more by bringing larger area under irrigation. Singh (1997) observed that under given water supply the area brought under irrigation in pearl millet was more in subnormal rainfall years than low rainfall years, however, the production enhanced in both the situations by bringing larger area under irrigation in pearl millet.

Deficit irrigation: Deficit irrigation approach seeks to avoid irrigation at less critical stages of crop growth and apply less water at the end so as to eliminate water stored at harvest. One has to be well versed with the crop growth stages less sensitive to moisture stress and proper balancing between water given and ET demand of crop. Shoot growth is very sensitive to mild soil moisture deficits and more sensitive than photosynthetic rate. Owing to this shoot growth sensitivity, it is common in many production systems to supply excess irrigation water to ensure rapid and extensive canopy development and thereby maximize radiation interception and biomass accumulation. There are three general scenarios for reducing water supplies to agriculture:

1. Season long drought management: a given volume is available for distribution within a fixed land area over the course of the growing season as the grower seems fit
2. Partial season drought management: a limited volume of water may be available over a fixed land area for only for a specific time period over a fixed land area
3. Spatial optimization strategies can be adopted such as moving production of specific crops to areas with greatest yield potential because of water availability or climatic and soil conditions

WATER HARVESTING

Water harvesting approaches is appropriate in dryland regions, by harvesting water on larger areas and channelling it onto crops. This is, in effect, a form of irrigation that has been practiced traditionally in some regions. The approach of water harvesting is capturing more water for crop transpiration. In arid and semi-arid parts rainfall varies from 150-350 mm and 350-550 mm respectively. Rainfall is not sufficient to ensure a good crop. In dryland farming, the solution to soil moisture problem lies in the storage of rainfall in the potential root zone of the soil by the water harvesting methods. Therefore, water harvesting would be an alternative to leave land fallow for increasing the available water supply for plant growth by way of -

(A) *In-situ* rainwater conservation: Water stored in soil profile itself using (1) Inter plot water harvesting, (2) Inter row water harvesting and (3) Pit and trench method for vegetable and fruits

(B) *Ex-situ* rainwater conservation: Water stored in reservoir or pond for recycling

Water harvesting is used to increase the total water supply available to crops. It is practiced in a variety of ways depending upon topography, soil and rainfall. Inter-plot water harvesting that uses a portion of land with 5% slope as catchment to generate runoff and divert it to adjacent area was found highly beneficial to improve the yield of many crops. CAZRI has developed water harvesting technique with 3 m cropped area and 1.5 m catchment area (catchment to crop area ratio = 0.5) with 5% slope on both sides (Singh 1988). Integration of regular row (RR) and double row (DR) plant geometries into water harvesting system on yield and water use efficiency of pearl millet best suited to runoff farming (Singh, 1988). Higher water storage in the 120 cm profile supported the pearl millet crop during ear emergence and grain development period thereby significantly increasing the yield attributing characters. Water harvesting proved well particularly in low rainfall years. The yield of pearl millet was almost three times with water harvesting compared to conventional sowing. In low rainfall situation in-situ water harvesting helped in making efficient use of limited rainfall and N-fertilizer in pearl millet (Singh 1985).

Micro-catchment systems are effective for use in growing trees and shrubs. In micro-catchment based cropping, rainwater is concentrated in a small portion of the cultivated area. The tree crops are deep rooted and can utilize the moisture stored in the sub-stratum and hence form a better option for micro-catchment based farming in sandy soil situations (Sharma et al 1986). Arid horticultural plants like pomegranate and ber can be successfully grown with appropriate micro-catchments in the water scarce regions. In micro-catchment systems, the catchment to collector area ratios may range from 1 : 1 to 20 : 1. The various sizes of micro-catchments have been used for different purposes. A number of catchment cropped area ratios and degree of slopes have been tried at CAZRI, Jodhpur. For ber, circular catchments of 1.5 m diameter and 5% inward slope with 54 m^2 of catchment has been found appropriate for conservation and proper utilization of rainwater (Sharma et al 1986). Increase in soil moisture by 20% was observed due to circular catchment or trenching than with any soil treatment (Ojasvi et al 1999). For further improvement in water use efficiency, these circular micro-catchments can be covered with plastic sheet (LDPE).

Ex-situ rainwater harvesting is a promising technology for enhancing the availability of water in arid areas. Ex-situ water harvesting involves collecting runoff originating from rainfall over a surface away from the field and storing it in surface storage systems for later use in the field. This type of rainwater harvesting provides supplemental or protective irrigation during dry periods of the cropping season.

IMPROVED IRRIGATION PRACTICES

Drip and sprinkler are the two examples of pressurized irrigation and both are able to save water as water is applied under pressure through a network of closed pipes, sprinkler nozzles and emitters. Sprinkler obtained significantly higher yield and WUE of wheat and potato. Sprinkler irrigation gained popularity in Sikar district of Rajasthan, as it is able to apply water on undulating type of topography. Other districts have also adopted this system of irrigation. But high wind velocity and saline water (>4 dSm^{-1}) restrict its utility in arid region (Singh et al 1978). Drip irrigation is the most efficient method of irrigating. While sprinkler systems are around 75-85% efficient, drip systems typically are 90% or higher. For this reason drip is the preferred method of irrigation in the desert regions. But drip irrigation has other benefits which make it useful almost anywhere. It is easy to install, easy to design, reduce disease problems associated with high levels of moisture on some plants but expensive. Drip irrigation works by applying water slowly, directly to the soil. The high efficiency of drip irrigation results from two primary factors. The first is that the water soaks into the soil before it can evaporate or run off. The second is that the water is only applied where it is needed rather than sprayed everywhere. The drip irrigation is neither affected by winds nor by saline water as it applies water directly in the rooting volume of crops. Studies conducted at CAZRI Jodhpur revealed that drip saves 30-50% water and provide 2-3 times higher yield than conventional irrigation. It is most suited to widely row spaced high value crops like vegetables, sugarcane, cotton, maize and plantation crops. With drip irrigation, a water with 3 dSm^{-1} was not so much detrimental to potatoes, while in tomatoes water up to 10 dSm^{-1} could be used successfully with drip irrigation (Singh et al 1978). Paired row planting system designed for each lateral minimized the installation cost and water use by 50%. This system is able to apply fertilizer with water, which ultimately led to higher yield of crops.

A variety of crops can be cultivated successfully with drip irrigation. These are: grapes, citrus, apples, pomegranate, pears, peaches, apricots, plums, bananas, dates, olives, mango, guava, tomato, green pepper, cucumber, lettuce, green pea, cauliflower, okra, cotton, sugarcane, corn, groundnut and onion, berries, melons, alfalfa, carnations, gladioli and roses.

Planting of seed or seedlings is done along the length of the lateral on one side or either side of it, irrespective of the position of the emitters on it. The emitters placed at a distance of 50 cm apart on a lateral line discharge the water at such rate that a continuous strip of about 40 cm width of soil remains wet throughout the length of the lateral. Different planting configurations (i) a single row (rectangular planting), (ii) double rows (square or equilateral planting) or (iii) triple row (hexagonal planting) were tried at CAZRI. The major objective of such different types of planting is to reduce the number of laterals, thereby reducing the cost of the system without reducing the plant population in the field. The results of the study revealed that equilateral planting with the side of 25 cm to 35 cm of a triangle performed the best in almost all the crops tried. Therefore, equilateral planting is advocated for vegetable crops like tomato, cabbage, cauliflower, turnip etc. (Singh and Singh 1978).

CONCLUSION

The problem of shortage of water to rainfed crops could be resolved by increasing water supply available to crops and increasing transpiration relative to evaporation losses. In case of limited irrigation resources, modest irrigation, deficit/extensive irrigation and pressurized irrigation exhibited the efficiency to improve water use and make efficient use of other resources like seed and fertilizer. Adoption of these approaches will certainly make efficient management of scarce water in arid and semi-arid region. Further, researches on these aspects in relation to area specific conditions are needed to provide proper package and practice to the end users.

REFERENCES

Daulay HS, Singh HP, Singh RP and Singh KC 1979 Effect of different mulches on yield and moisture use of pearl millet (*Pennisetum typhoides*). *Annals of Arid Zone* **18** 108-115

Gupta JP 1978 Evaporation from a sandy soil under mulches. *Annals of Arid Zone* **17** 287-290

Gupta JP and Gupta GN 1983 Effect of grass mulching on growth and yield of legumes. *Agricultural Water Management* **6** 375-383

Gupta JP 1980 Use of an asphalt subsurface barrier for improving the productivity of desert sandy soils. *Journal Arid Environment* **3** 215-222

Ojasvi PR, Goyal RK and Gupta JP 1999. The micro-catchment water harvesting technique for the plantation of jujube (*Ziziyphus mauritiana*) in an agroforestry system under arid conditions. *Agricultural Water Management* **41** 139-147

Saxena Anurag, Goyal RK, Singh RK and Roy MM 2014 Water Management for Climate Resilience in Arid Region. Central Arid Zone Research Institute, Jodhpur, India, 52 page

Saxena A, Singh DV and Joshi NL 1997 Effect of tillage and cropping systems on soil moisture balance and pearl millet yield. *Journal of Agronomy and Crop Science* **178** 251-257

Sharma KD, Pareek OP and Singh HP 1986 Micro-catchment water harvesting for raising jujube orchards in an arid climate. *Transactions of the American Society of Agricultural Engineers* **29(1)** 112-118

Singh GB 1985 Prospects of irrigation in arid zone. In *Irrigated Agriculture in Arid Areas* (ed) SD Singh, WAPCOS, New Delhi, 490-498 pages

Singh S D, JP Gupta and Panjab Singh 1978 Water Economy and Saline Water Use by Drip Irrigation. *Agronomy Journal* **70** 948-951

Singh SD 1978 Effects of Planting Configuration on Water Use and Economics of Drip Irrigation Systems. *Agronomy Journal* **70** 951-954

Singh SD 1988 *Water Harvesting in Arid Tropics*. CAZRI Publication No. 37, Central Arid Zone Research Institute, Jodhpur. 68 pages

Singh SD and Singh P 1978 Value of drip irrigation compared with conventional irrigation for vegetable production in a hot arid climate. *Agronomy Journal* **70** 945-947

Singh YV 1997 Maximising water use efficiency for sustainable crop production in arid ecosystem. In *Recent Advances in Management of arid Ecosystems* (eds) S Kathju, RK Abhichandani, Amal Kar and NL Joshi), Jodhpur, AZRAI, Jodhpur, 135-136 pages.

13

Development of drought tolerant crop varieties

R.K. Bhatt[1] and Ramavtar Sharma[2]

INTRODUCTION

Crop yields are normally affected by variations in climatic factors such as air temperature and precipitation and the frequency and severity of extreme events like droughts, floods and wind and hail storms. About 60% of cultivable area of India is rainfed or unirrigated. In the past India has experienced twenty four large scale droughts in 1891, 1896, 1899, 1905, 1911, 1915, 1918, 1920, 1941, 1951, 1965, 1966, 1972, 1974, 1979, 1982, 1986, 1987, 1988, 1999, 2000, 2002, 2009 and 2012 with increasing frequencies during the periods 1891-1920, 1965-1990 and 1999-2012. The challenge is thus to evolve well adapted self sustaining agricultural system for such labile conditions. The adaptation strategies would initially benefit from the technologies developed so far in the fragile arid ecosystem to combat drought and desertification, especially the improvements made in the traditional wisdom for water resources conservation and use, agroforestry systems and mixed cropping, etc. Efforts would then be required to enhance crop resilience and adaptation to climate change for sustainable production, food and nutrition security. This may be achieved by developing varieties of arid crops that are tolerant to flowering time high temperature, extra early varieties for shortened moisture availability period and varieties with terminal drought tolerance. Developing varieties with in-built developmental plasticity, which are early-maturing and capable of yielding grain under short wet periods but should have the ability to make best use of high moisture available during good rainfall years by prolonging their vegetative phase, resulting in higher grain and fodder yield. Concerted efforts are also required to understand the mechanism involved in controlling heat and terminal drought tolerance to identify genetic stocks for the traits correlated with these stresses.

---

1. Principal Scientist and Head, 2. Principal Scientist
Central Arid Zone Research Institute, Jodhpur (Raj.)
Email: rkbhatt@cazri.res.in

Developing the high yielding varieties for drought conditions with sustainable yield is a major goal of plant breeding. Earlier conventional breeding methods were used to select the genotypes for drought resistance/tolerance. In recent years, crop physiology and genomics have led to new insights in drought tolerance providing breeders with new knowledge and tools for plant improvement. Among the crops Pearl millet and sorghum are hardy dry land crops. Early maturing varieties of these crops have proved especially useful for dry land. Apart from millets, grain legumes are also important to restore soil fertility through nitrogen fixation. Cowpea, chickpea, lentil, moth bean, cluster bean, mung bean and pigeonpea are widely grown legumes under rainfed conditions in dry lands.

DROUGHT TOLERANCE ASSESSMENT

Most of the studies carried out on assessment of the degree of tolerance to drought in different crop varieties were based on yield and associated morpho-physiological characters. The difficulty in identifying a physiological parameter as a reliable indicator of yield in dry conditions has suggested that yield performance over a range of environments should be used as the main indicator for drought tolerance (Voltas et al 2005). Furthermore, a number of G X E (Genotype x Environment) interactions were also evaluated for genotype adaptability. Several indices were proposed to describe yield performance of a given genotype under stress and non-stress conditions or in comparison with the average yield or the yield of a superior genotype (Soika et al 1981, Lin and Binn 1988 and Yadav and Bhatnagar 2001). Some research workers considered canopy temperature and water potential for measuring the stress tolerance potential. Protective roles of antioxidants (superoxide dismutases (SODs), guaiacol peroxidase, catalase (CAT), ascorbate peroxidase (APX), and glutathione reductase (GR)) against oxidative damage under drought stress in plants and their help in minimizing the concentrations of reactive oxygen has been demonstrated.

PLANT STRATEGIES TO SURVIVE DROUGHT STRESS

Three strategies can help a crop to mitigate the effect of drought stress:

Drought escape: Ephemerals have a very short life cycle to escape drought ensuing seasonal showers. They come up with a few showers in desert ecosystem and complete their life cycle before moisture is depleted beyond sustainability. Production of large number of small seeds with long life further adds to their survival in the drier regions.

Drought avoidance: The drought avoidance is usually associated with perenniality or development of dormant structures for example rhizomes. Plants usually cease to grow and drop their leaves as a strategy to avoid draught. Resurrection plants on the other hand survive desiccation in an inactive dried state, without special dormant structures, which starts growing actively with the availability of moisture. For example *Tinospora cordifolia* (giloy) stem remains well hydrated even if it is placed under direct sun in summer where temperature rises beyond 45°C and starts rooting and sprouting during its growing season or with increased moisture level in the air.

Drought tolerance: Tolerance to desiccation is usually achieved through short term adjustment of metabolic processes to survive concomitant dry spell. The underlying mechanism might be similar to the one for drought avoidance but such plants are usually not equipped with special features to survive extreme desiccation. Interplay of a number of genes is required to maintain and protect cell functionality and structure during the stress.

PLANT DROUGHT TOLERANCE MECHANISMS

Plants respond to their changing environment by effecting multi-functional and structural changes. Therefore, the genetic control of tolerance to abiotic stresses is not only very complex, but is also highly influenced by other environmental factors and developmental stage of the plant. The physiological responses of plants to deficit of water include leaf wilting, a reduction in leaf area, leaf abscission, and the stimulation of root growth by directing nutrients to the underground parts of the plants. In addition, increased level of abscisic acid (ABA), a plant stress hormone, induces the closure of leaf stomata thereby reducing water loss through transpiration, and decreasing the rate of photosynthesis to improve the water-use efficiency. Matching the whole plant adaptation plant cells start accumulation of solutes such as, sugars, amino acids, organic acids and ions – especially potassium (K^+) in order to maintain water balance and thereby normal cellular function.

Regardless of strategies adopted the plant response to drought is accompanied by the activation of genes involved in the perception of drought stress and in the transmission of the stress signal. At least four independent stress responsive genetic regulatory pathways are known to exist in plants, forming a highly complex and redundant gene network. A group of genes encoding structural proteins is then activated to protect the cells from the effects of desiccation. These genes include those that govern the: accumulation of compatible solutes; passive transport across membranes; energy-requiring water transport systems; and protection and stabilization of cell structures from desiccation and damage by reactive oxygen species. Drought tolerance mechanism is genetically controlled and genes or QTL responsible for drought tolerance have been discovered in several crops which opens avenue for molecular breeding for drought tolerance.

DROUGHT TOLERANCE TRAITS

Osmotic adjustment: As a plant experiences a water-deficit stress, it accumulates a variety of osmotically active compounds such as amino acids, sugars and ions inside its cells, resulting in a lowering of the cell osmotic potential. Water present in inter-cellular spaces then flows into the cells. This process, called "osmotic adjustment" (OA), was proposed as a potential factor that could enable plants to maintain turgor and survive better at low water status.

Cell membrane stability (CMS): The ability to maintain the cell wall pressure to maintain the integrity of cell under dehydration. This is the tolerance potential of cell membrane to prevent electrolyte leakage at decreasing water content of crop varieties and therefore cell activity is maintained. Measuring of CMS has been used in different crop varieties for selecting the tolerance.

Chlorophyll stability index: Stress induced reactive oxygen species causes damage to chlorophyll and photosynthetic apparatus, the genotypes that maintain the chlorophyll content of the chloroplast are usually found to yield better and also the chlorophyll stability has been considered as selectable trait for heat tolerance of genotypes.

Partitioning and stem reserve mobilization: Under drought conditions the photosynthesis is decreased consequently, grain filling depends on the reserve material of stem. Numerous studies have reported that stem reserve mobilization capacity is related to yield under water-stress in wheat. In rice, a few studies also indicated that this mechanism maintains grain yield under water stress at the grain filling stage. This drought tolerance mechanism is stimulated by a decrease in gibberellic acid concentration and an increase in abscisic acid concentration.

IMPROVEMENTS FOR DROUGHT TOLERANCE

High yielding abiotic and biotic stress tolerant varieties are developed through different approaches of plant breeding either using breeding approaches or genetic engineering. Breeding for any desired trait undoubtedly requires a significant amount of genetic variation at intra-specific, inter-specific or inter-generic levels. The available variability, may further be enhanced by new gene combinations, intercrossing those genotypes that hold desirable characteristics and introducing new germplasm from other existing breeding programs (Flowers 2004 and Ashraf and Akram 2009). The induced mutagenesis could also be a potent technique in creating newer allelic variability.

Conventional breeding for drought tolerance

`Conventional breeding approaches are widely used for development of genotypes for drought tolerance since long time. Conventional breeders at different national and international research centers have made considerable efforts in developing drought tolerant lines/cultivars of some important food crops. For example, breeding approach started at the International Maize and Wheat Improvement Center (CIMMYT), Mexico in the 1970s for developing drought tolerant maize is worth mentioning.

Marker-assisted breeding (MAB) for drought tolerance

Through marker-assisted breeding it is now possible to examine the usefulness of genomic regions of a crop germplasm under water limited regimes, which was, in fact, previously not possible. Drought tolerance is controlled by many minor genes (polygenes) that have additive effects in their expression (Zhao, 2002, Mohammadi et al 2005). Thus, the loci on chromosomes housing such types of genes are now referred to as quantitative trait loci (QTL). Natural genetic variation of a crop can be exploited either via direct selection under stressful conditions whether simulated or natural or via mapping of QTL (polygenes) and subsequent marker-assisted selection (Ashraf et al 2008). QTL mapping allows assessing the locations, numbers, magnitude of phenotypic effects, and pattern of gene action (Vinh and Paterson 2005). The role of polygenes in controlling a trait has been widely assessed by traditional means, but the use of DNA markers and QTL mapping has made it convenient to dissect the complex traits (Humphreys and Humphreys 2005).

For a QTL analysis, phenotypic evaluation is carried out of a large number of plants from a population segregating for a variety of genetic markers; then a part or the whole population is genotyped; and finally linkage is established to determine the loci controlling a trait (Asins 2002). Due to the intricacy of abiotic stress tolerance and the problems encountered in phenotypic based selection, the QTL mapping has been considered as imperative to the use of DNA markers for improving stress tolerance (Ashraf et al 2008). DNA markers such as RFLPs, RAPDs, CAPS, AFLPs, microsatellites (SSRs), SNPs, and DNA sequences are used to assess the inheritance of stress tolerance. QTL mapping for the drought tolerance trait has been done in different crops, the most notable being maize, wheat, barley, cotton, sorghum, and rice (Saranga et al 2001, Sanchez et al 2002 and Bernier et al 2008). QTL involved in variation in relative water content (RWC), number of tillers (NL) and total shoot fresh mass were located on a RFLP-RAPD genetic map. It was also found that different DNA regions mediate in constitutive water stress responses. In view of a number of previous reports, it is evident that molecular mapping and a number of QTL associated with drought tolerance identified in different crops can be effectively used in appropriate breeding programs meant for improving crop drought tolerance. QTL mapping is thus an effective way of detecting specific components that permit direct appraisal of stress tolerance.

With the development of detailed molecular linkage maps, a significant progress has been made during the last few years in marker-assisted selection procedures that have allowed pyramiding of desirable traits to attain substantial improvements in crop drought tolerance. For example, Steele et al (2006) conducted a marker-assisted back-crossing (MABC) breeding program to improve some key root morphological characteristics, involved in the drought tolerance of an Indian upland elite rice variety, Kalinga III.

Early maturity, high drought tolerance and high grain yield with good grain quality are the prominent characteristics for breeding a novel variety. At IRRI, efforts have also been made to improve drought tolerance of rice using the marker- assisted breeding approach.

Pearl millet [*Pennisetum glaucum* (L.) R. Br.] is known for its high drought tolerance, however, drought tolerance being independent of yield ability it could not be converted into yield stability and drought still remains a major constraint for its optimum production in many areas of the world. Thus, breeding for productivity under water stress in pearl millet is a major challenge for many research institutes. The breeding research at ICRISAT, India, has resulted in mapping several QTL for stover and grain yield under terminal water deficit conditions (Serraj et al 2004). In one of the study with pearl millet, line 863B has been found to have a superior general combining ability for the grain filling trait under terminal drought conditions (Yadav et al 2004). A mapping population developed from a cross 863B × ICMB 841 was evaluated under early- and late-drought stress conditions, which led to the identification of two genomic regions in line 863B associated with improved panicle harvest index as well as with high drought tolerance (Yadav et al 2004).

Transgenic approach to enhanced drought tolerance

To meet out the food security all over the world, plant scientists worked towards gene revolution after green revolution due to advances in biotechnology. The gene revolution, in fact, involves modification of qualitative and quantitative traits in an organism by transferring desired genes from one species to another. This strategy is referred to as the transgenic approach. Transgenic approach allows the incorporation of only the specific cloned genes into an organism and restricts the transfer of undesirable genes from donor organism. Through this approach, pyramiding of genes with similar effects can also be achieved. Rapid advance in recombinant-DNA technology and development of precise and efficient gene-transfer protocols have resulted in efficient transformation and generation of transgenic lines in a number of crop species (Gosal et al 2009).

Transgenic approach is being pursued actively throughout the world to improve traits including tolerance to biotic and abiotic stresses in a number of crops (Ashraf et al 2008). As with salt stress, plant responses to drought stress are complex, because it involves many genes with additive effects, so the prospects of improving drought tolerance in crops seem not to be very direct. Despite this, efforts have been made during the last few decades to generate transgenic lines of different crops, which have shown improved tolerance to drought stress. The major emphasis of bioengineers has been on engineering genes that encode compatible organic osmolytes, plant growth regulators, antioxidants, heat-shock and late embryogenesis abundant proteins, and transcription factors involved in gene expression.

MAJOR AGRO-CLIMATIC ZONES IN RAJASTHAN AND ITS MAJOR CROPS

Rajasthan is the largest state in the Indian Union with a geographical area of 3.42 lac sq. km. About 77 percent of its population lives in rural areas and about 70 percent depend on agriculture as source of livelihood. The region to the west and north-west, comprising eleven districts and nearly 61 percent of the area of the state is known as the great Indian Thar Desert. The soil type in this region is sandy which is poor in nitrogen and has low water holding capacity. The rainfall fluctuates from less than100 mm to 400 mm in the western arid parts and 500 mm to more than 900 mm per annum in south-eastern and eastern parts. The Rajasthan state is classified in to ten major agro-climatic zones based on the annual rainfall pattern, soil types and various agricultural situations met with and these are briefly described in Table 1. The crop distribution and cropping pattern has largely evolved around available moisture and rainfall pattern providing sustainability to the farming system.

Suitable dryland crop varieties, which match with the rainfall pattern of the region and efficiently utilize the rainfall and stored soil moisture, have been developed. Mutation breeding, particularly through gamma rays has proved successful in delivering a number of short duration varieties in moth bean (Henry et al 1993, Kumar 1996). Many improved mutants which vary for their tolerance to drought (Soni et al 2011) have been developed at CAZRI, Jodhpur (CZM-1, CZM2) and SK Rajasthan Agricultural University, Bikaner (RMO-40, FMM-96, RMO-225, RMO-257and RMO-435) (Sharma and Kakani 2002). Short duration varieties with good yield ability are also available in mung bean, cowpea and pearl millet.

Table 1. Major agroclimatic zones of Rajasthan

Zone	Districts	Rainfall	Major crops
Ia	Barmer, part of Jodhpur	200-370	Pearl millet, guar, moth, sesame
Ib	Ganganagar, Hanumangarh	100-350	Cotton, guar
Ic	Bikaner, Jaisalmer, Churu	100-350	Pearl millet, guar, mothbean
IIa	Nagaur, Sikar, Jhunjhunun, parts of Churu	300-500	Pearl millet, guar, pulses
IIb	Jalore, Pali, parts of Sirohi and Jodhupur	300-500	Pearl millet, guar, sesame
IIIa	Jaipur, Ajmer, Dausa, Tonk	300-500	Pearl millet, guar, sorghum
IIIb	Alwar, Dholpur, Bharatpur, Sawai Madhopur, Karauli	500-700	Pearl millet, guar, ground nut
IVa	Bhilwara. Sirohi, Udaipur, Chittorgarh	500-900	Maize, pulses, Sorghum
IVb	Dungarpur, Udaipur, Banswara, Pratapgarh	500-1100	Maize, paddy, sorghum, Blackgram, Mungbean
V	Kota, Jhalawar, Bundi, Baran	650-1000	Sorghum, soybean

Table 2. Dominant grass species of arid zone and their habitat

Species	Rainfall (mm)	Habitat	DMY (t ha^{-1})
Lasiurus sindicus	200-250	Sandy soils	5 - 7.5
Cenchrus ciliaris	300-400	Well drained soils	4 - 4.5
C. setigerus	300-400	Well drained soils	0.5 – 2.5
Panicum antidotale and *P. turgidum*	> 250	Sandy to sandy loam	2 - 5
Dichanthium annulatum	> 380	Heavy soils	3 -5
Sehima nervosum	> 380	Hilly terrain, shallow	3-5

PASTURE GRASSES

About 106 species of grasses are found in western Rajasthan. The important grass species of Rajasthan are listed in Table 2. The grassland cover of Indian arid zone with particular reference to western part of Rajasthan is of *Dichanthium-Cenchrus-Lasiurus* type (Buffel, dhaman, sewan). At the Central Arid Zone Research Institute, Jodhpur a sizable genetic stock of about 255 collections derived from different habitats is being maintained. Sewan (*Lasiurus sindicus* Henr.) is called as 'King' of desert grasses which survives under extreme arid conditions in a rainfall range of 100 -300 mm. Buffel grass or Anjan grass (*Cenchrus ciliaris* L.) is adapted to a wide range of soil and climatic conditions and can be cultivated in areas receiving rainfall from 150 to 1250 mm. Bird Wood Grass or Dhaman (*Cenchrus setigerus* Vahl) is tolerant to heat and drought and grows in areas of annual rainfall as low as 200 mm. Karad (*Dichanthium annulatum* (Forsk) Stapf) occurs mainly in the rainfall regions of ≥ 350 mm. Blue Panic (*Panicum antidotale* Retz.) often found on sand dunes and works as an excellent sand binder and drought resistant perennial grass adapted to a variety of soils and climatic conditions.

Dominant grasses of arid region

- *Lasiurus sindicus, Panicum turgidum, Panicum antidotale, Cymbopogon jwarancusa, C. scheonanthus, C. parkerii, Eleusine compressa, Dactyloctenium sindicum, D. aegypticum, Cenchrus setigerus* and *Cenchrus ciliaris.*
- Annual grasses: *Aristida adscensionis, Cenchrus biflorus, Eragrostis tremula, E. tenella* and *Tragus biflorus.*

FUTURE PERSPECTIVES

Development of novel crop varieties for tolerance to abiotic stresses particularly for terminal drought and heat tolerance is highly expected to cope with the unfavourable environmental challenges. Although lot of work and information generated to develop the drought tolerant varieties but still information is needed to manipulate the complex quantitative traits. Apart from the conventional and traditional approach for developing the drought tolerant varieties new approaches like, marker assisted breeding and transgenic engineering is needed to develop the improved high yielding and stable crop varieties.

REFERENCES

Ashraf M and Akram NA 2009 Improving salinity tolerance of plants through conventional breeding and genetic engineering: an analytical comparison. *Biotechnological Advances* **27** 744–752

Ashraf M, Athar HR, Harris PJC, Kwon TR 2008 Some prospective strategies for improving crop salt tolerance. *Advances in Agronomy* **97** 45–110

Asins MJ 2002 Present and future of quantitative trait locus analysis. *Plant Breeding* **121** 281—291

Bernier J, Kumar A, Serraj R, Spaner D, Atlin G 2008 Review: breeding upland rice for drought resistance. *Journal of Science, Food & Agriculture* **88** 927–939

Flowers TJ 2004 Improving crop salt tolerance. *Journal of Experimental Botany* **55** 307–319

Gosal SS, Wani SH, Kang MS 2009 Biotechnology and drought tolerance. *Journal of Crop Improvement* **23** 19–54

Henry A, Daulay HS and Bhati TK 1993 A promising moth bean variety for arid and semi arid regions. *Indian Farming* **43** 21-23

Humphreys MO and Humphreys MW 2005 Breeding for stress resistance: general principles, *Abiotic stresses: plant resistance through breeding and molecular approaches;* Agronomy *and crop production* (eds) M Ashraf and PJC Harris, 19–46 pages

Kumar D 1996 Variability studies in induced mutants of moth bean on rainfed arid lands. *Annals of Arid Zone* **35** 125-128

Lin CS and Binn MR 1988 A superiority measure of cultivar performance for cultivar × location data. *Canadian Journal of Plant Sciences* **68** 193–198

Mohammadi M, Taleei A, Zeinali H, Naghavi MR, Ceccarelli S, Baum M Grando 2005 QTL analysis for phenologic traits in doubled haploid population of barley. *International Journal of Agricultural Biology* **7(5)** 820–823

Sanchez AC, Subudhi PK, Rosenow DT, Nguyen HT 2002. Mapping QTLs associated with drought resistance in sorghum (*Sorghum bicolor* L. Moench). *Plant Molecular Biology* **48** 713–726

Saranga Y, Menz M, Jiang CX, Wright RJ, Yakir D, Paterson AH 2001 Genomic dissection of genotype × environment interactions conferring adaptation of cotton to arid conditions. *Genome Research* **11** 1988–1995

Serraj R, Krishnamurthy L, Kashiwagi J, Kumar J, Chandra S, Crouch JH 2004 Variation in root traits of chickpea (*Cicer arietinum* L.) grown under terminal drought. *Field Crops Research* **88** 115–127

Sharma RC and Kakani R 2002 *In*: *Moth bean in India* (eds) D Kumar and NB Singh, Scientific Publishers (India), Jodhpur, 9-30 pages

Soika RE, Stolzy LH, Fischer RA 1981 Seasonal drought response of selected wheat cultivars. *Agronomy Journal* **73** 838–845

Soni P, Rizwan M, Bhatt KV, Mohapatra T and Singh G 2011 *In-vitro* response of *Vigna aconitifolia* to drought stress induced by PEG-6000. *Journal of Stress Physiology & Biochemistry* **7(3)** 109-121

Steele KA, Price AH, Shashidar HE, Witcombe JR 2006 Marker-assisted selection to introgress rice QTLs controlling root traits into an Indian upland rice variety. *Theoretical Applied Genetics* 112 208–221

Lang NT, Chi Buu B 2008. Fine mapping for drought tolerance in rice (*Oryza sativa* L.). *Omonrice* **17** 22-28

Salvi RS 2006. Genomics-based approaches to improve drought tolerance of crops. *Trends in Plant Science* **11** 405-412

Vinh NT, Paterson AH 2005 Genome mapping and its implication for stress resistance in plants. *Abiotic stresses: plant resistance through breeding and molecular approaches* (eds) M Ashraf and PJC Harris

Voltas J, Lopez-Corcoles H, Borras G 2005 Use of biplot analysis and factorial regression for the investigation of superior genotypes in multi-environment trials. *European Journal of Agronomy* **22** 309–324

Yadav OP and Bhatnagar SK 2001 Evaluation of indices for identification of pearl millet cultivars adapted to stress and non-stress conditions. *Field Crop Research* **70** 201–208

Yadav RS, Hash CT, Bidinger FR, Devos KM, Howarth CJ 2004 Genomic regions associated with grain yield and aspects of post-flowering drought tolerance in pearl millet across stress environments and testers background. *Euphytica* **136** 265–277

Zhao J 2002 QTLs for oil content and their relationships to other agronomic traits in an European × Chinese oilseed rape population, *Diss. Grorg-Agust University of Goettingen, Germany.*

14

Contingency Crop Planning for Drought Management

T.K. Bhati

INTRODUCTION

Drought is a recurring phenomenon in western Rajasthan occurring once in three years to alternate year. Its intensity also varies from district to district e.g., 43% of the years in Jaisalmer, 30% of Barmer, 17% of Jodhpur faced severe drought in last 100 years, whereas the corresponding values of moderate drought were 27, 20 and 28 per cent in these districts. However, the impact of drought also depends on several factors like quantum and period of water scarcity, prevalent crop and their varieties, cropping system, management and strategies adopted, governmental support and policies etc.

To cushion the adverse effects of weather aberrations, farmers in western Rajasthan are following crop diversification through mixed sowing of dryland crops in varying proportion depending on edapho-climatic situations and family requirement. The practice is now getting lesser preference due to development of high yielding varieties (HYV) and improved production technology. However, as an alternative to mixed cropping, various R&D institutes have developed new approaches viz., area allocation under different crops and their varieties as per climatic and social considerations, improved cropping systems and sequences, introduction of new crops, etc. which if properly dovetailed with suitable management practices can help to mitigate the adverse effect of droughts and weather aberration. Also with the use of weather advisory services provided by Indian Meteorological Department (IMD), the crop planning and management practices can further be refined and timely implemented to combat drought. Nevertheless, importance of perennial components like pasture grasses, MPTS, medicinal and high value herbs and shrubs can't be undermined in this region. Therefore, they should find proper place in crop diversification models not only

Principal Scientist (Retd.)
Central Arid Zone Research Institute, Jodhpur (Raj.)
Email: bhatitej@gmail.com

to effectively face the extremities of drought but also for sustainability of agriculture. The important considerations with respect to management of crops and cropping systems under varying drought conditions are discussed.

CONTINGENCY CROP PLANNING AS PER ONSET OF MONSOON

The normal date of sowing rains (> 25 mm week^{-1}) in western Rajasthan are between 1st to 15th July but the onset of monsoon in the region is very erratic. Therefore, the selection of crops and their proportion will depend on varied Agro-ecological situations (AES) and other socio-economic factors, for example if the onset of monsoon is timely the diversification of crops is recommended in the ratio 4:3:2:1 pearl millet, pulses, clusterbean and sesame respectively in the situations of 300 to 400 mm rainfall having light to medium textured soils. However, the crop diversification should be adopted with suitable crop varieties (short and medium duration) and recommended production technology.

The anticipated monsoon is divided into four periods i.e. 20-25th July, 26-31st July, 1-6th August and 7-15th August and a contingency plan is drawn. The approach is to sow short duration pulses/cereal/oilseed and also to produce fodder, which will receive priority in case of further delay of monsoon

Table 1. Contingency crop plans as per onset of monsoon

Contingency operation	Possible period of monsoon arrival			
	20-25 July	**26 - 31 July**	**1 - 6 August**	**7 - 15 August**
Field preparation	Field bunding, removal of bushes and harrowing across the slope			
Crop & varieties	Green gram (RMG 62, RMG 268, RMG 344, SML 668, GM 4)	Moth bean (CZM-2, RMO 40, RMO 225, RMO 257, RMO 435)	Moth bean (CZM-2, RMO 40, RMO 257, RMO 435)	Moth bean (CZM-2, RMO 40, RMO 435)
	Moth bean (RMO 40, RMO 257, RMO 435, CZM 2)	Clusterbean (RGC 936, RGC 1002, RGM 112, RGC 1003, RGC 1017)	Clusterbean (RGC 936, RGC 1002, RGM 112, RGC 1003)	Clusterbean for fodder (RGC 936, RGC 1017)
	Clusterbean (RGC 936, RGC 1002, RGC 1017, RGC 1003, RGM 112)	Greengram (RMG 62, RMG 268, RMG 344, SML 668, GM-4)	-	-
	Cowpea (RC 19, RS-9, C-152, RS-9, C-152, V-16, JC-5, JC-10)	Pearlmillet for fodder (Raj 171, MH 169) with cowpea (Bundel Lobia-1, UPC 9202, IC421, FS68)	Pearlmillet for fodder (Raj 171, MH 169) with cowpea (Charodi-1, Bundel Lobia-1, UPC 9202, IC421)	Pearlmillet for fodder (Raj 171, MH 169) with cowpea (Bundel Lobia-1, UPC 9202, IC421)
	Sesamum (RT 46, RT 125, RT127, RT 346 with moth bean/ mung bean in inter cropping system in 2:2 ratio)	Sesamum RT46 with moth bean as mixed crop		

	Pearlmillet, extra-early to early maturing varieties (HHB 67, HHB 67-2, ICMH 356, RHB 121, RCB-2, GHB 538) Inter cropping Pearl millet (HHB 67, HHB 67(+), ICMH 356, RHB 121, GHB 538) with Greengram (RMG 62, RMG 268, RMG 435)	Pearl millet for fodder (Raj 171, MH 169) with Mothbean (RMO 40, RMO 257, RMO 435, CZM 2)	Pearlmillet for fodder (Raj 171, MH 169) with Mothbean (RMO 40, RMO 257, RMO 435, CZM 2)	
Pre-treatment of seeds for drought resistance and crop growth	Soak the seeds in potassium nitrate or common salt (7.5-10 g per 10 L water or 1.5-2 tea spoons per 10 L bucket of water) at night for 2 hours, then dry it for 2-3 hours, before sowing Thiourea, if available, @5g per 10 L water, or 1 tea spoon in 10 L bucket of water for 2 hours; then dry for 2-3 hours and use for sowing.			
Sowing	In furrows at 60 cm row to row spacing for grains and broadcasting for fodder			
Seed rate	Normal rate	Increase seed rate by 10-15% of normal (fodder crop)	Increase seed rate by 10-15% of normal (fodder crop)	Increase seed rate by 10-15% of normal (fodder crop)
Fertilizer	For greengram, Moth bean and Cowpea Basal N:P (10:20 kg/ha) For pearl millet Basal (20 kg N/ha); top dressing (20 kg N/ha) depending upon rain	For Greengram, Moth bean and Cowpea Basal N:P (10:20 kg/ha) For pearl millet No basal; top dressing (15-20 kg N/ha)	For Greengram, Moth bean and cowpea No fertilizer For pearl millet No basal; top dressing (15-20 kg N/ha)	For Greengram, Moth bean and cowpea No fertilizer For pearl millet No fertilizer
Insect pest and disease management	Greengram, Moth bean and Cowpea should be treated with 2 gram carbendazim per kg of seeds to check dry root rot or treat the seed with *Trichoderma* culture @ 4 g per kg of seeds. Clusterbean seeds should be soaked in 0.025% Streptocycline solution (1 g in 4 litre water) for 2 hours to check bacterial blight. To check termite attack, Chloropyrophos 1.5% dust should be mixed in soil @ 25 kg per hectare at the time of sowing for all crops.			

Mid-season Corrections for Crops already Sown

Long dry spell after sowing cause poor germination and death of seedlings . Under such situations field should be kept free from weeds and dust or vegetative mulch should be applied. Besides, if rains commence, transplanting of pearlmillet can also be done. Mid season drought occurs when there is long dry spell for consecutive 2 weeks rainless (>2.5 mm) period or more either at vegetative or reproductive stage. Following measures should be carried out under such circumstances.

- Timely weed control and interculture operations.
- Mulching/ crust breaking.
- Reduction of plant population - removal of alternate rows in case of drought (if the rows are originally spaced at 30 cm)
- Spray of thiourea at 500ppm
- Proper plant protection measures
- Lifesaving irrigation to the crop as per availability and stage of crop

Contingency Crop Planning in Runoff Farming System (*Khadin*)

In western Rajasthan runoff farming systems broadly exists in three different situations. The crop contingency planning for these situations is given below:

Situation I: Monsoon occurrence is normal with respect of frequency and quantum: Under such situations depending on the runoff and water storage status sowing of *kharif/rabi/kharif-rabi* crops may be planned.

Situation II: Monsoon occurrence is timely but quantum is very low: Go for sowing of *kharif* crops viz. pearl millet and arid legumes in the part of the *khadin* where rain water is not standing but sufficient soil moisture is available for crop production.

Situation III: Monsoon occurrence is delayed by 15-20 days but the quantum of runoff is high: In upper part of *khadin* go for sowing of short duration varieties of *kharif* crops and in lower part of *khadin* after removal of excess water go for cultivation of *rabi* crops like wheat, barley, chickpea, mustard, taramira, cumin, isabagol etc.

Important considerations for management of *rabi* crops in runoff farming systems

- Short duration varieties of *rabi* crops need to be selected for example Raj 4037, Raj 3077, Raj-1482, Raj 3777, of wheat; RSG-44, RSG-888 of chickpea; T-59, RS-30, Ashirwad, RH-819, Bio-902 of mustard; RD-31, BL-2, RD-2052, DL-157, DL-36 of barley, RTM314 of taramira, RZ19, GC4, RZ209 of cumin, GI2, RI89, RI1 of isabgol.
- Before sowing, seed should be treated with proper fungicide, insecticide and microbial culture.
- Proper doses of nutrients should be used in the crops as per soil test, availability of soil moisture and the crop under cultivation through organic and inorganic sources.
- Improved production technology with respect to plant geometry and cropping systems, *in-situ* moisture conservation and weed management, control of insect pest & diseases, protection of the crops from frost and

hot winds. should be adopted. Further in *khadins* surface and ground water utilization facilities (water harvesting structures, dug well, tube well), to save the crops from total failure /to improve to the productivity irrigations at critical crop growth stages should be given.

Long-Term Strategies for Drought Mitigation

- Rain water harvesting and its recycling through sprinklers/drips for sustainable crop production.
- Development of common property resources viz., grazing lands, orans, nadis etc. and their efficient management to reduce animal pressure on arable crops.
- Land use planning on watershed/Index catchment/*Khadin* basis with inclusion of arable crops, grasses, multipurpose tree species and animals in farming system mode. The systems could be agri-horticulture with ber and gonda; agro-forestry with *Prosopis cinearia*, *Tecomella undulata*, *Acacia senegal*; silvi-pasture (grasses viz. *Cenchrus ciliaris, Lasiurus sindicus* and fodder shrubs and trees viz. *Zyziphus nummularia*, *Hardwickia binnata, Z. rotundifolia*) for sustainable agriculture production which is also conducive for drought management in annual crops.
- To evolve drought resistant and high yielding varieties of cereals, legumes, oil seeds and fodder crops.
- To popularize cultivation of drought hardy and highly remunerative traditional medicinal and high value plants like *Guggal, Asvagandha, Shatawari, Tumba, Sonamukhi* etc. as new crop introductions for increased income and better drought sustainability of arable lands.
- Development of suitable infra-structure for marketing of value added products made out of crop biomass to help generate alternative income opportunities to the farmers.
- Harnessing of solar and wind energy for utilization in agriculture.

CONCLUSION

Hence, it is concluded that the adverse effect of drought on crop production in western Rajasthan can be substantially reduced with the adoption of proper crop contingency plans prepared as per short, medium and long term weather forecasting and executed taking into account the recommended crops, varieties, cropping systems and the production technology. However, for sustainability of crop production in context to the changing scenario of climate in this tract, long term drought management strategies are needed to be implemented through governmental interventions *vis a vis* community participation by way of adoption of proper policy framework, infrastructure development and institution building.

15

Eco-Technologies for Drought Mitigation

Arun K. Sharma

INTRODUCTION

Climate-related natural disasters such as drought, floods, storms and tropical cyclones are the principal sources of risk and uncertainty in agriculture. The wide fluctuations in agricultural output that have occurred throughout human history attest to the fact that agriculture is an economic activity dependent on the vagaries of weather. As a result, climatic variations significantly impact the livelihood of millions – and particularly poor and vulnerable people. Drought is and will remain a major determinant of agriculture in the low rainfall regions. Farmers make various adjustments in their crop production, consumption and livelihood practices and adopt conservative measures to reduce the negative impacts of drought. The production adjustments and conservative measures reduce direct production losses. The consumption and livelihood adjustments, on the other hand, help in smoothing consumption costs and income. Nevertheless, these adjustments themselves entail some economic costs in terms of lost opportunities for income gains in good years. More importantly, the appropriateness and effectiveness of adjustments depend on the information, knowledge and skills of individual farmers and communities.

The economic costs of drought include not only the crop production loss, but also the loss in production of post-monsoon crops grown on residual soil moisture. Production practices designed to reduce losses from drought cause additional economic and social costs, as do the depletion of productive assets due to distress sale during drought years and the consequent long-term decline in productive capacity. Drought thus causes enormous socio-economic and environmental impact in terms of food and livelihood insecurity, increasing

---

Senior Scientist
Central Arid Zone Research Institute, Jodhpur (Raj.)
Email: arun.k_sharma@yahoo.co.in

poverty and vulnerability, and negative environmental consequences. It is believed that the frequency and severity of drought are increasing due to climate change. As a result, the socio-economic impact is expected to rise in the future. The effect of drought on human societies can be multidimensional. Severe droughts can result in starvation and death of the affected population. However, different types of economic costs arise before such severe consequences occur. As a result of market failures, farmers attempt to 'self-insure' by making costly adjustments in their production practices and adopting conservative measures to reduce the negative impact in drought years. In rural areas where agricultural production is the major source of income and employment, a decrease in agricultural production will set off second-round effects through forward and backward linkages of agriculture to other sectors.

NEED OF ECO-TECHNOLOGIES

In drought prone areas farmers due to weak socio-economic conditions are not able to adopt costly measure for drought mitigation and in such situation they need locally available low cost technologies for drought mitigation. These technologies should also be eco-friendly and help in making a support system for drought mitigation. Since water is the most limiting factor during drought eco-technologies that manage water will be effective in drought mitigation.

In India about 41% agriculture area receives rainfall less than 500 mm/year. These areas are mainly in Rajasthan, Haryana, Gujarat, Andhra Pradesh, Karnataka, Maharashtra and Tamilnadu states. Low rainfall coupled with its high variability, wide range of annual temperature (0°C to 49°C) and light soils make agriculture a challenge in these areas. Many a times crops are either not able to complete life cycle or yield poorly due to this complex interplay of climatic conditions. This variability has been further aggravated due to climate change effects as evidenced by trend of climatic data and experienced by the farmers in these areas in the last decade (Rao 2009). Agriculture systems with annuals (crops) and perennial (trees, shrubs and grasses) has been followed traditionally in these areas to spread the risk of climatic uncertainty and to get food, fodder and fuel for sustenance of life. In the last 4-5 decades several attempts have been made to improve productivity of this traditional system with the use of external inputs (synthetic) e.g. fertilizers, pesticides, weedicides but success is limited to the good rainfall year as these inputs perform better with assured water availability, rather than during below average rainfall years and use of these synthetic inputs give negative results too.

In low rainfall areas due to high risk associated with crop production, the use of agro chemicals is very less. Thus farmers are practicing eco-friendly/ organic agriculture since ages by default. The system despite its low productivity compared to conventional system has established itself as highly efficient in terms of resource recycling and providing better food and economic sustainability in the arid region (Sharma and Tewari 2009). The improvement in this production system is possible by nutrient management and plant protection aspects. For nutrient management practice of ley farming, inclusion of legume in crop rotations and efficient recycling of biomass are important for the region. For effective non-chemical plant protection some measures e.g. crop rotation (Lodha 2008), use of compost as soil application along with foliar spray of compost extract (Lodha and Burman 2000), application of oil cakes or residues of mustard with one summer irrigation (Lodha 2008) and application of neem based pesticides (Satyavir and Yadav 2005) have been found promising.

TECHNOLOGIES FOR COLLECTION AND STORAGE OF RAINWATER

These technologies help in conserving the limited available water. In this group all such measures are included that are used for collecting maximum of rainwater and *in situ* moisture conservation.

Collection of rainwater

To minimize the impact of drought, soil needs to capture the rainwater that falls on it, store as much of that water as possible for future plant use, and allow for plant roots to penetrate and proliferate. Problems with or constraints on one or several of these conditions cause soil moisture to be one of the main limiting factors for crop growth. Bunding around the field is most appropriate technology for this purpose. On the bunds trees and grasses should be planted so that bunds remain intact. Some of the frequently flowering plants e.g. senna should also be planted on bunds that not only gives some economic produce during drought but also enhances faunal diversity helpful for pollination and pest control. Bunding has been the age old practice to reduce runoff and improve moisture storage in the soil. Bunding prevented runoff, increased infiltration and improved the availability of moisture to *rabi* crops i.e., mustard, taramira and chickpea by 11.4 mm in a soil core of 100 cm (Regar et al 2007). Inter-plot water harvesting and collection of rainfall in farm pond for life saving irrigation can be another option for collection of rainwater.

Increasing soil moisture storage capacity

This can be possible by three ways:

(i) Increasing water infiltration;

(ii) Reducing soil evaporation; and

(iii) Increasing soil moisture storage capacities.

All the three are related to soil organic matter content. For increasing water holding capacity addition of organic manure is the most appropriate method. For every 1% of organic matter content, the soil can hold 16,500 litres of plant-available water per acre of soil down upto 30 cm depth. Thus, the amount of organic matter in a soil directly influences the availability of water to a crop over time. In order to create a drought-resistant soil, it is necessary to understand this most important factor influencing soil moisture. Increased levels of organic matter and associated soil fauna lead to greater pore space with the immediate result that water infiltrates more readily and can be held in the soil. Bioturbating activities of earthworms and other macro-organisms and channels left in the soil by decayed plant roots improve pore space. The consequence of increased water infiltration combined with a higher organic matter content is increased soil storage of water. Organic matter contributes to the stability of soil aggregates and pores through the bonding or adhesive properties of organic materials, such as bacterial waste products, organic gels, fungal hyphae and worm secretions and casts. Moreover, organic matter intimately mixed with mineral soil materials has a considerable influence on increasing moisture holding capacity, especially in the topsoil. This can be achieved by application of manure at the time of sowing and manual weeding. The use of mulch and cover crops, not only saves water but also adds organic matter in the soil.

Any application of animal manure, slurry or other carbon-rich wastes improves the organic matter content of the soil. In some cases, it is better to allow a period of decomposition before application to the field. Well matured compost made with animal and crop waste is the most appropriate technology for enhancing organic matter content in the soil. In arid regions where the organic matter level of soil is very low and soil structure is very weak, soil amendments like pond sediments, vermiculite, FYM etc. were found very promising in improving moisture retention capacity of soil. Pond sediments available from widely scattered ponds are commonly utilized for improving soil productivity. The mixing of pond silt upto 30-40 cm soil depth @ 76 t ha^{-1} increased available water storage capacity from 6.5-6.9%, reduced infiltration rate from 15 to 13.2 cm h^{-1} and hence increased the yield of pearl millet by 40-50% and greengram by 35-40% over control. Since a long time farmyard manure, has been known to maintain physico-chemical properties of soil and widely used by the farmers. Surface application of FYM and Milkweed (*Calotropis procera*) increased the yield of pearl millet by 30% (Agarwal and Sharma, 1980). The consequence of increased water infiltration combined with a higher organic matter content is increased storage of water in soil.

TECHNOLOGIES FOR EFFICIENT USE OF HARVESTED RAINWATER

Cover crops/legumes

Growing cover crops is one of the best practices for improving organic matter levels and, hence, soil quality. The benefits of growing cover crops are as follows:

- Prevents erosion by anchoring soil and lessening the impact of raindrops.
- Adds plant material to the soil for organic matter replenishment.
- Some cover crops bind excess nutrients in the soil and prevent leaching.
- Provide habitat for beneficial insects and other organisms.
- Moderates soil temperatures and, hence, protects soil organisms.

Leguminous crops enrich the soil with N and their residues decompose rapidly because of their low C:N ratio. Later, when the system is stabilized, it is possible to include cover crops with an economic function, e.g. livestock fodder.

Microbial inoculants

Different types of microbial inoculants like *Rhizobium*, *Azotobactor*, phosphate solubilizing bacteria are available and increase the crop yield with a very little investment. The nitrogen-fixing bacteria are present in the Indian arid soils but they vary with increase and decrease in organic matter content in the soil. *Rhizobium* inoculation alone in greengram obviates the needs of N and P under natural conditions in arid agriculture (Singh 1977). Microbial properties of arid soils have large variation with their land use pattern, soil moisture status and soil organic matter content (Rao and Vankateswarlu 1983). *Azospirillum brasilense* inoculation to pearl millet led to increase of 37.9% in tillers, 44.3% in heads and 32.1% in test weight of pearl millet (Joshi and Rao 1989). The inoculation effect was reduced with increase in N application. Soils of dry lands are considered not only thirsty but also hungry. Crops grown in dry land may not respond to the fertilizer application in some cases where moisture availability is poor. Integrated nutrient management also has its own restrictions in rainfed condition. Bio-fertilizer has a scope of fertilizing the crop through augmenting the nutrient supply or their enhancement in availability to crop. Most of the nitrogen-fixing bacteria relevant to the rainfed cropping are legume-*Rhizobium* symbiosis, for example, groundnut and soybean. Most of the cultivated soils in India are reported to have relatively large population (>100 g-1 dry soil) of *Rhizobia* capable of nodulating the legumes.

Mulching

Covering soil with weeds biomass and crop residues not only reduce moisture losses from soil but also add organic matter in long run. A combination of tillage and mulches can benefit soil organic matter content, bulk density and erodibility properties besides improved soil and moisture conservation (Bhatnagar and Sur 1984). Mulch affects the soil life. Litter placement on the soil surface (as opposed to incorporation with ploughing) increased the ratio of fungi to bacteria - the reason being that fungi have a higher carbon assimilation efficiency than bacteria. In addition, it encourages bioturbating (mixing) effects of macrofauna bringing the materials in surface layers of the soil. The results obtained in different locations clearly indicated that crops gave better response when tillage especially minimum/zero tillage was combined with surface residues or mulches.

When adequate crop residues are available, conservation tillage is highly effective for conserving soil and water; achieving favourable crop yields, maintaining soil organic carbon contents and soil and water quality. However, when crop residues are limited or removed for some other purposes, other tillage methods, namely, contouring, tide ridges, graded furrows terracing etc., along with appropriate conservation practices may be needed. Leaving surface residue is key to control runoff, soil erosion and hard setting in rainfed areas, which are the key problems. In view of the shortage of residue in rainfed areas, in arid regions, several alternative strategies have emerged for generation of residue either through *in situ* cultivation and incorporation as a cover crop or harvesting from perennial plants grown on bunds and adding the green leaves as manure cum mulching. Agroforestry and alley cropping systems are other options where biomass generation can be integrated along with crop production

Manual weeding

Weeds use a large proportion of water if not controlled. As a drought mitigation measure weeds must be removed by intercultural operation and should be placed on the soil surface as mulch. Intercultural also breaks capillary movement of soil water. Frequent weeding is an important part of dryland agriculture. Line sowing and mechanical weeding, with appropriate size of blade harrows, remove unwanted vegetation which competes with the main crop. Weeding within rows can be done using hand hoes. Removal of unwanted vegetation helps the main crop to have greater accessibility to soil moisture and plant nutrients for its own growth. Thus by following this technology a great loss of water can be checked and that helps a lot in drought mitigation.

Intercropping/mixed cropping

Inter/mixed cropping system is the best way to minimize the risk and sustain the production in drought prone areas. Mixed cropping of different crops along with the main crops, such as millets and different legumes, is an insurance against the vagaries of the monsoon. The different root systems of mixed crop feed at different depths of the soil. Moreover, mix cropping provides small quantities of grain of different kinds for home consumption at different times. The intercropping increases the yield and stabilizes production not only by increasing water and nutrient/fertilizer use efficiency but also due to minimising the competition for space and improved light energy conversion (Joshi 1999). Intercropping suppresses weeds better than sole cropping and thus provides an opportunity to utilize the crops themselves as a tool of weed management. With every care taken to undertake timely agricultural operations, it is still possible that the whole operation becomes a gamble due to unpredictable monsoons. The main crop could fall in the early part of its life cycle. In such cases, the farmer should come up with an alternate crop that can mature in a very short time and under hard conditions to take advantage of what is left of the rainy season. Contingent planning helps catch and make the best use of late rains. Advance planning is necessary in selecting a contingent crop. And all the requisites for its sowing should be ready within the main season itself.

Crop geometry

Crop geometry is distribution and arrangement of crop plants in the field. It determines the area available to each plant and the competition for inputs mainly moisture, nutrients, light and space. In arid region, particularly under rainfed condition, larger canopy may be disadvantageous as it may exhaust the available soil moisture from root zone during the drought condition (Singh 1977). Therefore, in the drought prone areas, enough considerations should be given for maintaining optimal plant stand and row spacing. While studying effect of intra row spacing and time of thinning in pearl millet hybrids, it was observed that thinning at three week stage keeping 15-20 cm intra row spacing (at recommended row spacing of 45 cm) is a beneficial practice (ARS 1990).

Diversified production system

Instead of monoculture, multiple crops/ trees/shrubs need to be grown in a given area or field. This spreads risk of drought and farmers get some income from any one or more produce of the system during drought. The demand of spices and medicinal plant is increasing when grown under ecofriendly management, so it must be promoted as a component of production system in the various agro-climatic regions (Table 1) of the drylands (Sharma 2011).

Table 1. Potential high-value crops for organic agriculture in different agro-climatic regions

Agro climatic zone	Rainfall Average (mm)	Suitable species and medicinal crop
Arid	100-250	Cumin, Senna, Gugool, Psyllium, Clusterbean (gum)
Transitional	250-350	Fenugreek, Cumin, Aloe, Senna
Semi arid	350-500	Fennel, Asvagandha, Fenugreek, Coriander, Lucorice, Aloe, Aonla

Besides adding organic matter to the system, perennial trees and shrubs recycle plant nutrients from deeper soil layers through their rooting system. Through litter and pruning, these can be used again by annual crops. Probably the most important contribution of perennials in a production system lies in the fact that throughout the whole year their roots excrete root exudates and decaying root cells, which in turn are used as an energy source by soil microorganisms. The food web in the soil is maintained, even during dry seasons when no annual crops are grown. The result is that soil biota are in place to provide the crop with nutrients at the beginning of the next cropping season.

Integrated pest management

As with manuring, proper pest and disease management results in healthy crops. Healthy crops produce optimal biomass, which is necessary for organic matter production in the soil. Diversified cropping and mixed crop-livestock systems enhance biological control of pests and diseases through species interactions. Through integrated production and pest management farmers learn how to maintain a healthy environment for their crops. They learn to examine their crops regularly in order to observe ratios of pests to natural enemies (beneficial predators) and cases of damage, and on that basis to make decisions as to whether it is necessary to use natural treatments (using ecofriendly products such as neem or other botanicals) and the required applications.

INTEGRATED USE OF ECO TECHNOLOGIES IN MODEL ORGANIC FARM AT CAZRI FOR MITIGATING DROUGHT

A two hectare certified model organic farm (MOF) was established during 2008, within the central research farm CAZRI, Jodhpur. Following Eco-technologies in collection and storage of rainwater have been integrated at the MOF to develop environment for research on organic system-

- A trench cum mound is made around the farm for *in-situ* conservation of rainwater and to avoid contamination through water that might flow in from adjoining fields. *Cassia angustifola,* a medicinal shrub is planted on the mound for round the year availability of flower for predators and further filtration of contamination.
- Two rainwater harvesting tanks of 5000 liters capacity (each) are constructed in the model organic farm. Cemented catchment area is made around each pond for maximum collection of rainwater. This catchment area is also utilized for drying and thrashing of crops during the lean period. The rain water so collected is used for raising low volume-high value crops like cumin, psyllium etc.
- Manual weeding is done regularly and uprooted weeds are spread in the same field as mulch that got decomposed in due course and contributed 1.5-2.0 t/ha of organic matter
- Agro-diversity plantation has also been done, that includes fruit trees namely *Ziziphus mauritiana, Phyllanthus emblica, Cordia myxa,* shrubs e.g. henna (*Lawsonia alba)* for shelter to predators, and for biopesticides *Adathoda vasica, Vitex nigundo, Aloe vera* and *Casia angustifolia*. Besides these, naturally grown, 30 plants of Khejri (*Prosopis cineraria,* a climax MPTS of the region) and 2 plants of neem (*A. indica)* are also

protected. This plantation ensured adequate nectar supply and shelter to the beneficial insects. Neem plantation has been done around the field for preparation of bio pesticides.

- Six compost pits are made at one corner of the farm for making compost from the MOF generated crop waste.
- Biopesticides are prepared with leaves of neem, calotropis, adhatoda etc. and neem cake.
- Pheromone traps are installed for trapping the pests like while grub, moth etc.

Crop resilience to high climatic variability e.g drought, excessive rains was enhanced at model organic farm, with the use of organic manure along with the above mentioned eco-technologies as is evident from the two years data. Since crop yield is the ultimate product of all the interactions therefore only yield comparison is given here.

- During 2009 because of low (208mm) as well as highly erratic (only 4 good rainy days and 2 long dry spell of 17 and 29 days) rainfall, crop faced drought conditions. The severity was further increased by dry-hot winds during milking stage of crop in September. With this set of conditions while crops in the conventional plots were failed, but the crop grown with organic inputs coupled with timely interculture for moisture conservation yielded up to 30-40% of the average yields. A yield of 196.3 kg/ha of sesame and 206.8 kg/ha of cluster bean was obtained in sampled plots (Table 2).
- In 2010, above normal rainfall (460 mm) during the cropping season was recorded. Under such situations organic plots due to organic matter incorporation maintained optimum aeration and moisture in soil that resulted in good crop growth while in the conventional plots crop growth was stunted and gave poor yield (Table 2).

Table 2. Crop resilience to climatic variability (average rainfall at experimental site is 366 mm/yr)

Year and rainfall	**Crop yield (Kg/ha)**			
	Sesame		**Cluster bean**	
	Organic	**Conventional**	**Organic**	**Conventional**
2009 (rainfall 208 mm)	196.3	0.0	206.8	0.0
2010 (rainfall 460mm)	886.6	523.9	630.2	308.6

Pest Management : Pests can be kept below economic threshold level with the integrated use of following eco- technologies –

- Soil application of neem seed powder @ 300 kg/ha

- Use of well prepared compost @ 5.0 t/ha
- *Trichoderma viride* application in soil and seed treatment
- Use of healthy seed and also free from weed seeds
- Hand weeding and mulching
- Prophylactic spray of neem seed kernel extract solution(5.0%) at regular interval
- Use of pheromone traps according to the pest
- Regular visit of every part of the field to assess crop health conditions and timely spray was done at the initial stage of pest attack.

CONCLUSION

Eco-technologies are cost effective, easy to apply and improve environment of the drought prone areas. Through experiments these have been proved helpful for mitigation of drought to a great extent. The only need is their timely use to get full benefits and minimize the effect of drought.

REFERENCES

Agarwal RK and Sharma VK 1980 Availability of nitrogen and phosphorus during the decomposition of *Calotropis procera* (Aak) in desert sandy soil. *Journal of Indian Society of Soil Science* **28** 407-409

ARS 1990 *Kharif Report*. Agricultural Research Station, Mandore, Jodhpur. pp. 24-44.

Bhatnagar VK and Sur HS 1984 Role of tillage and crop residue in controlling water erosion and research needs. *Journal of Indian Society of Soil Science* **12** 43-53.

Joshi NL and Rao AV 1989 Response of pearl millet to *Azospirillum* as influenced by nitrogen fertilizer under field condition. *Annals of Arid Zone* **28** 291-297

Joshi NL 1999 Effect of component densities on the performance of pearl millet- mung bean intercropping system under different rainfall situations in arid environment. *Annals of Arid Zone* **38(1)** 15-25

Lodha S 2008 Organic amendments and biocontrol agents for plant disease management. In *Organic Agriculture* (eds.) JC Tarafdar, KP Tripathi and M Kumar, Jodhpur: Scientific Publishers, 200-218 pages

Lodha S and Burman U 2000 Efficacy of composts on nitrogen fixation, dry root rot and yield of legumes. *Indian Journal of Agricultural Sciences* **70** 846- 849

Rao AS 2009 Climatic variability and crop production in arid western Rajasthan. *In Trends in Arid Zone Research in* India (eds) Amal Kar, B.K. Garg, M.P. Singh and S. Kathju, CAZRI, Jodhpur,48-61pages

Rao AV and Venkaseswarlu B 1983 Microbial ecology of the soils of Indian desert agriculture. *Agriculture Ecosystems and Environment* **10** 361-364

Regar PL, Rao SS and Joshi NL 2007 Effect of *in-situ* moisture conservation practices on productivity of Rainfed Indian mustard (*Brassica juncea*). *Indian Journal of Agronomy* 52**(2)** 148-150

Satyavir and Yadav SS 2005 Managing insect pest complex in arid legumes: an eco-friendly approach. *Journal of Arid Legumes* **2** 357-360

Sharma AK and Tewari JC 2009 Improvement in Traditional Agroforestry Systems with Organic Inputs in Arid Zone. *In Conservation Farming* (eds) S. Bhan and R.L. Karle, New Delhi: Soil Conservation Society of India, 292-295 pages

Sharma AK 2011 Action plan for organic seed spices production. *In Recent advances in seed spices* (eds) Y Rabindrababu, RK Jaiman and KD Patel, New Delhi:. Daya Publishing House, 85-93

Singh SD 1977 Effect of rhizobia inoculation on nodulation and yield of mung bean (*Vigna radiata* L. Wilczek.). *Annals of Arid Zone* **6** 79-84.

16

Drought Proofing through Dryland Horticulture

P.R. Meghwal[1] and Akath Singh[2]

INTRODUCTION

Drought is a many faceted natural disaster that leads to serious socio-economic impact particularly affecting agricultural production and water supplies. There are two distinct phases in which the application of knowledge of weather and climate change can reduce the impact of drought on communities. The first is long term planning in which strategies can be devised, and precautions taken to reduce the impact. The second phase is the action taken during the onset of the event to reduce adverse effect. The efforts are on to stabilize dryland agriculture especially by involving perennial drought hardy fruit plants in contingent crop production strategies in rain fed areas of India. Arid and semi arid regions by and large suffer from the vagaries of the nature. The rainfall is low, unpredictable, uncertain, erratic and insufficient. The extremes of temperature, high wind velocity, problematic soils and deep ground water make the situation further worse. The dryland horticultural crop if exploited properly can play significant role in alleviating the adverse effects of drought. Following strategies may be adapted to increase the resilience of dryland horticultural plants for drought proofing.

DELINEATION OF CROPS AND VARIETIES

Looking to agro climatic-edaphic situation of the region, specific crops and varieties have been identified. These include several indigenous and other commercial horticultural species (Table 1, Fig. 1).

Many of fruits and vegetable crops mentioned above are indigenous and have capacity to withstand drought and salinity and still give good economic returns.

---

1. Principal Scientist, 2. Senior Scientist
Central Arid Zone Research Institute, Jodhpur - 342 003 (Rajasthan), India
Email: meghwal_pr@yahoo.in

More drought resistant species and varieties/genotypes have been identified which give satisfactory yield under low rainfall and with climatic adversities. Among the arid fruit crops, ber, pomegranate, aonla, kair, lasoda, drumsticks, khejri, kummat etc. have the ability to perform even under limited rainfall and tolerate several abiotic stresses. Locally adapted *Ziziphus* species like *Z. nummularia* and *Z. mauritiana* variety *rotundifolia* are resistant to frost and supply multiple products such as fodder, fruits and fuel wood.

Fig. 1. Fruit trees of arid zone

Table 1. Potential horticultural crops for arid regions

Fruits	Vegetables
Ber (*Ziziphus mauritiana*)	Water melon/Mateera (*Citrullus lanatus)*
Ker (*Capparis decidua)*	Musk melons (*Cucumis sativus)*
Lasoda (*Cordia myxa)*	Bottle gourd *(Lagenaria siceraria*)
Kumat (*Acacia senegal)*	Round melon (*Citrllus lanatus* var. *fistulosus*)
Tamarind (*Tamarindus indica*)	Long melon (*Cucumis melo* var. *utillismus)*
Bordi (*Ziziphus rotundifolia*)	Snap melon (*Cucumis melo* var. *momordica*)
Phalsa (*Grewia subinaequalis)*	Bittergourd *(Momordica charantia*)
Pilu (*Salvadora oleoides*)	Kachari (*Cucumis callosus*)
Khejri (*Prosopis cineraria*)	Cucumber (*cucumis sativus)*
Aonla (*Emblica officinalis)*	Ridge gourd *(Luffa acutangula*)
Jamun (*Syzygium cumini)*	Sponge gourd (*Luffa cylindrica*)
Bael (*Aegle marmelos*)	Cow pea (*Vigna unguiculata*)
Jhar ber (*Ziziphus nummularia*)	Clusterbean (*Cyamopsis tetragonoloba)*
Karonda (*Carissa carandas*)	Amaranthus (*Amarnathus* spp.)
Natal plum(*Cariss grandiflora*)	Brinjal (*Solanum melongena*)
Prickly pear (*Opuntia ficus-indica*)	Chillies (*Capsicum annum)*
Date palm (*Phoenix dactylifera*)	Tomato (*Lycopersicon esculentum*)
Drumstick (*Moringa oleoides*)	Carrot *(Daucus carota*)
Fig (*Ficus carica)*	Onion (*Allium cepa*)
Papaya (*Carica papaya)*	Garlic (*Allium sativa*)
Guava (*Psidium guajava*)	Radish (*Raphanus sativus*)
Grape (*Vitis vinifera*)	Okra (*Abelmoscus esculentus*)
Custard apple (*Annona squamosa*)	Cabbage (*Brassica oleracia* var. *capitata*)
Sour lime (*Citrus aurantifolia*)	Cauliflower (*Brassica oleracia* var. *botrytis*)
Kinnow (*Citrus* spp)	Potato (*Solanum tuberosum*)
Sweet orange (*Citrus sinensis)*	Pea (*Pisum sativum* var. *hortans*)

COLLECTION, CONSERVATION AND EVALUATION OF GERMPLASM FOR DROUGHT PROOFING

Wide range of germplasm of different fruit crops have been collected, evaluated, conserved and suitable varieties identified for facing the challenges of drought and its possible ill effects in future (Table 2). It is also imperative to continue collection of diverse germplasm of different horticultural plants as a source of gene pool for drought and other abiotic stress resistance. Long term evaluation of

22 ber cultivars under different rainfall situations indicated that varieties CAZRI Gola, Gola, Sanaur-5 and Tikadi showed better yield even in low rainfall years. The performance of ber in providing security against recurring drought has been such that it is designated as an insurance against drought. It has inbuilt mechanism to evade water stress since it enters in dormancy during peak summer and restart vegetative growth with the onset of monsoon rain. Most of the early varieties of ber complete flowering, fruiting and fruit harvest before soil moisture recede. Even the late varieties can be managed with a few supplementary irrigations from harvested rain water or from some other source in rainfed areas. Similarly, early varieties have been identified in aonla, date palm etc. Profitable crop of lasora fruits can be managed with little irrigation during flowering and fruiting time while for the rest of the season it can survive without irrigation once established.

PROPAGATION AND NURSERY MANAGEMENT

The role of plant propagation and nursery management cannot be undermined in tackling the problem of drought. If the planting material remains available, large number of plants can be planted whenever rainfall is abundant. Poly tube method of raising rootstock seedlings and getting the buddable seedling in 90 days enables the availability of budded ber plants in the month of July for making the best use of early rain for better establishment of the plants in the first year. This technique was adapted by a number of private nurserymen and millions of budded ber plants have been produced and supplied throughout the country. This helped enormously in increasing the acreage under ber to about 90,000 hectares in India. The use of nitrogen fixing bacteria, AM-fungi and their combination had significant influence on growth and nutrient uptake of ber seedlings besides producing healthy seedlings for budding 10-20 days earlier than uninoculated ones (Meghwal et al 2006). The treatment of lasora (*Cordia myxa*) seeds with gibberellic acid (250 & 500 ppm) promoted seed germination which was further improved upon rubbing the stem end of seeds with sand paper followed by GA_3 treatment (Meghwal 2007). In karonda, highest seed germination was recorded under seed treatment with GA_3 @ 200ppm . Similarly in aonla, seed germination could be improved significantly by treatment with 1% KNO_3. In Kair (*Capparis decidua*), hardwood cutting rooted better (30-40%) in July-August with no significant influence of IBA on rooting per cent (Meghwal and Vashishtha 1998). In pomegranate, treatment of hard wood cutting with 1000 ppm IBA for 30 seconds gave 60-80% success in rooting (Bankar and Prasad 1989). In custard apple, T-budding resulted in 70-75% success in the month of March (Bankar 1989). In lasora, I-budding during July gave highest success in bud take but budding with equally good rate of success could be done up to September after which it started declining (Meghwal 2007).

Table 2. Dryland horticultural crops and their suitable varieties

Name of fruit species	No. of varieties	Accessions	Recommended varieties
Ber	40	-	Gola, Seb, Mundia, Umran, Tikadi, Illaichi, Banarsi Karaka, Chhuhara etc.
Pomegranate	24	-	Jalore Seedless, Ganesh, G-137, Mridula
Aonla	8	-	Kanchan, Chakaiya, NA- 7, Anand-2, Krishna
Date palm	18	-	Barhee, Halawy, Muscat-2, Khadrawy, Medjool, Khunaizi, Zahidi etc.
Gonda (lasora)	-	20	CAZRI-G-2021, CAZRI-G-2025
Karonda	1	6	Pant Manohar, CZK-2011, CZK2022,CZK-2031
Bael	4	4	Dhara Road, Faizabadi Local, NB-5, NB-9
Guava	2	--	Allahbad Safeda, L-49
Sour Lime	2	2	Kagzi lime,Vikram, Pramalini etc
Kair	-	7	-
Fig	-	1	Black Ischia, Puna , Dinkar, Dienna
Phalsa	-	2	-
Khirni	-	2	-
Prickly pear	-	4	-
Khejri		3	Thar Shobha
Mullberry		2	-
Ivy gourd	4	-	-

SELECTION OF SUITABLE ROOTSTOCK

Rootstocks play important roles in utilization of natural resources. A plant's first line of defence against abiotic stress is in its roots, which highlights the importance of rootstocks in fruit production. Drought hardy rootstocks are very important for sustaining the plants in case of drought or other abiotic stresses. Rootstock studies on ber revealed bordi as the best rootstock for commercial ber varietues Gola, Seb, Umran etc. (Anonymous, 1989). The rootstocks in grapes, citrus, ber, and in many temperate fruit crops play important roles in enhancing water use efficiency, mitigating the adverse impact of biotic and abiotic stresses, scion compatibility, canopy architecture, fruit quality and nutrient absorption. In case of ber and lasora for example bordi (*Ziziphus mauritiana* var. rotundifolia) and goondi (*Cordia gharaf*) respectively have been identified as resistant rootstock against moisture stresses and frost. Use of salt and drought tolerant rootstocks provides a viable option for sustainable grape production under saline conditions. Selection of rootstocks with high nutrient uptake efficiency will decrease production cost and contribute to fruit production in citrus and grapes. Grape rootstocks such as *Vitis berlandieri* and *V. rupestris* were considered as drought tolerant, while Dogridge and 110R as salinity resistant, hence, judicious use of rootstocks in different tropical and sub-tropical fruit crops are helpful in achieving sustainable productivity, reducing the input cost, combating stresses and efficient utilization of problem soils. In sapota (*Achras japota*), khirni (*Manilkara hexandra*) has proved to be drought tolerant and productive rootstock in marginal soils. Citrus rootstock, Rangpur lime can adapt to alkalinity, salinity and calcareous soils and can impart resistance to *Phytopthora,* cold hardiness besides high quality fruits.

TRAINING AND PRUNING

Proper training and pruning are essential operations for successful horticulture crop production. Training is performed during initial years to develop a strong framework of scaffold branches to bear the crop load and for equal distribution of sun light and air. This is also important to keep the orchard free from diseases and pests as proper aeration and Sun light distribution in entire canopy areas help to minimize pest and disease incidence. Pruning is essentially done for fruiting in some fruit crops such as ber and phalsa while in others it is limited to removal of diseased and dried branches. Light pruning is recommended under drought and low rain fall conditions. During the pruning operation, the unproductive part of the past season's main shoot and its secondaries as well as weak and diseased cris-cross branches are removed to obtain healthy growth at most of the productive nodes. The pruning of water sprouts and other undesirable branches saves necessary metabolites which are diverted for the growth of desirable shoots and thus enhance the overall efficiency of the production under limited moisture and nutrient supply.

WATER MANAGEMENT

Water is the most precious input for dryland horticulture and therefore, its use should be planned judiciously. Rain water harvesting and its management play significant role in horticultural production. This is a practice of converting more rain water into soil water. The rain water can either be diverted directly to tree basin *in situ* or in suitable structures *ex situ* which can later be utilized as lifesaving irrigation. In sandy soils, *in situ* conservation while in heavier soils *ex situ* conservation should be popularized (Reddy 1999). The water thus collected remains stored deep into soil profile, escapes from evaporative losses and is available during critical period of demand. Perennial fruit trees once established get the advantage of any time rains during the years unlike annual crops. Water harvesting in *khadins* in parts of Jaisalmer district of Rajasthan and its use in raising of crops and trees is the age old practice of *in situ* water harvesting. The hilly, rocky and sloppy lands generate run off which can be collected in pond or tanka and recycled for establishment of fruit plantations and lifesaving irrigation at critical stages. The selection of bahar and irrigation scheduling accordingly assumes great significance in most of the dryland fruits. In pomegranate and ber for example, mrig bahar utilises rain water effectively and regular irrigation need to be given only for a limited period from September onwards till February after which it is withheld to regulate cropping. Ber trees, once established can give some production under rain fed condition depending upon rainfall, but provision of irrigation water increases the yield significantly. Work done at CAZRI revealed that fruit yield per tree varied from 10 - 52 kg in rain fed condition with 125-800 mm rainfall, respectively. Higher run-off yield was obtained from catchment having 5% slopes (Sharma et al 1982). Circular catchments around each tree (1.5 m radius) with 5% slope towards tree trunk and covering the catchment with black polyethylene sheet has been advocated to concentrate and conserve the moisture. If there is enough rain during July-August months, irrigation is not required during this period. After fruit set, irrigation at 15 days interval from November to February is beneficial for fruit growth and development under north Indian conditions. The application of irrigation water at 1.0 IW: CPE ratio at 10 cm depth with 11 to 12 irrigation from June to February was found to be the best for pomegranate for higher fruit yield (Anonymous 1990-91).

The studies on drip irrigation revealed early commercial fruit production in ber, pomegranate and aonla with the application of water equal to 60% of pan evaporation (PE) and also registering significantly higher fruit yield compared to 40% PE.(Anonymous 2007). In kagzi lime, drip irrigation at 0.7 Etc was better over 1.0 Etc in terms of increasing plant growth and saving of water under Bikaner conditions (Beniwal et al 2006). The spatial distribution of salts was minimum at 20 cm lateral spacing in all the irrigation levels. Results from a study conducted on chilli at CAZRI suggested that it is better to schedule irrigation at 0.8 of E-pan evaporation and apply on canopy area basis combined with 180 kg ha^{-1} nitrogen maximize the production per unit of water. A significant interaction between water and nitrogen was found. Daily drip fertigation maintains high soil moisture and nutrient in available form in the root zone which might be responsible for higher chilli production (Singh et al 1999). Drip irrigation increased yield of long gourd by 45 to 47%, round gourd by 21 to 38% and water melon by 10 to 22% compared with sprinkler and furrow irrigation under coarse textured soil of arid region. The water use efficiency was nearly twice with drip irrigation as compared to other methods of irrigation (Singh and Singh 1978). Drip fertigation provided gainful use of nutrient management with water and provided maximum yield of tomatoes and chilli (Singh and Kumar 1999, Singh et al 1997). It has been observed that daily drip irrigation maintained higher moisture content and subjected the salts beyond active root zone thereby minimizing the detrimental effects of salinity (Gupta and Singh 1983). Poor quality water (10 dSm^{-1}) could also be used through drip in tomatoes without much deterioration in yield and quality (Singh et al 1978).

Mulching with organic material like straw, hay, dry leaves, local weeds etc. has been found to increase the availability of water and its content in soil by decreasing evaporative losses to alleviate harmful effects of drought. This practice also suppresses weed growth, prevents soil erosion and adds to organic matter build up to the soil. Black polythene mulch has been found to be very effective in ber orchard. Organic mulches like grasses, weeds and crop residues reduced soil temperature considerably during summer months, while plastic films raised soil temperature during winter months (Srivastava et al 1984).) Among the various organic mulches, fruit yield per plant in brinjal was highest in plots mulched with lasora (*Cordia myxa*) leaf followed by kheep (*Leptadenia species*) clippings (Awasthi et al 2006). Use of lasora leaf as mulch increased ascorbic acid, acidity, B-Carotene and vitamin A content of brinjal fruits followed by white and black polyethylene mulches. The water loss due to transpiration can be reduced by use of radiation reflectants, stomatal closing chemicals and plastic films. Spraying of 4-6% Kaolin and 0.5-1.0% liquid paraffin considerably reduces plant water loss. Chemicals such as phenyl mercuric acetate, abscisic acid and acetyl alcohol cause stomata closure and thereby reduce transpiration. Provisions of wind breaks also modify microclimatic situations leading to reduction in transpiration loss. For moisture conservation, field bunding, mulching, organic manure application, use of fly ash (15%) exfoliated vermiculite and bentonite clay was found effective in crop lands (Faroda 1998). Application of exfoliated vermiculite in vegetable crops under arid condition have been found to increase the moisture retention capacity at 0.1 bar tension and decreased the bulk density and saturated hydraulic conductivity of soil. Its application at the rate of 20 t ha^{-1} optimized the production of lady's finger which was 40-60% more than the control without vermiculite (Gupta and Gupta 1982). Bentonite clay was found to be an effective sub-surface barrier in reducing soil moisture losses due to deep percolation which showed significant yield increase in round gourd (Singh et al 1979).

NUTRIENT MANAGEMENT

The arid zone soils are very poor in majority of the nutrients except potash. Enrichment of soil with organic and inorganic sources is essential for taking optimum production from fruits and vegetables. Application of chemical fertilizer have made miracle initially in agriculture and horticulture. However the increasing awareness of the environmental degradation, due to various anthropogenic activities has been expressed world over about the over reliance on mineral fertilizers causing environmental penalties. It has also led to deficiency of several micronutrients. Therefore, the option left is whether to revert back to traditional farming using organic input or to follow some middle path by integrating both organic and inorganic. So far as present knowledge is concerned, organic manuring alone cannot maintain the required productivity and demand of the food. Integrated nutrient management by inclusion of organic, inorganic and biological sources of nutrients seems to be the best way to sustain the productivity without impairing soil quality. For an efficient integrated nutrient management, various components are to be used in a balanced form to make horticultural system more profitable on sustainable basis.

INCREASING FRUIT SET DURING DROUGHT

Under the overall influence of drought, the decreased vegetative cover leads to temperature rise and dry atmosphere (less humid environment). Such environment diminishes the diversity and number of pollinator insects which in turn affect the fruit set, cause more fruit to drop that sets and affect the final fruits retention. The microclimate in the orchard at flowering and fruit set is important for fruit set. It can be improved by planting shelterbelts. The population of pollinator insects will increase upon the improvement in the surrounding environment. Fixing of bee hives in orchard can also enhance pollination. Considerable research work is required to identify the specific horticultural management practices that can increase pollination and thus the yield.

MANAGING BIOTIC STRESSES

Pest and disease occurrence under arid ecosystem is comparatively less than other regions such as humid tropics. Integrated approach for control of pest and diseases has been suggested to tackle the problem. The basic idea behind this approach is to reduce the use of chemical pesticides and to promote the application of cultural practice and bio pesticides. Termite is the number one enemy for most fruit crops in low rain fall areas. The increase in termite population and consequent attack on dryland horticultural crops increase many folds during drought/moisture stress. Hence, special care and precautions are needed to manage this pest during drought. Clean cultivation and regular inspection of the orchard help to detect the attack of pest. Environmentally safe management practices such as cultural, mechanical/physical and biological methods for the control of biotic stresses in dryland horticultural crops may be useful in managing the biotic stresses without harming the environment and with low cost.

TOP WORKING

Seedling plants of bordi (*Ziziphus mauritiana* var. *rotundifolia*) and jhar ber (*Ziziphus nummularia*) are of common occurrence throughout arid and semi-arid regions of India. Traditionally, these plants are being used as fodder, fuel wood, timber or as fencing material in rural areas. Similarly seedling plants of aonla, khejri, bael etc. are also found in different areas. These plants may be converted into promising varieties by top working with scion shoots from improved varieties of respective crops (Fig. 2). Studies conducted by CAZRI at farmers field reveals that *in situ* budded ber can start commercial fruiting in second year under rain fed conditions.

Fig 2. Top working in ber (left); *in situ* budding in khejri (right)

POST-HARVEST MANAGEMENT

The losses in post-harvest are quite high in India i.e. 25-30 per cent which could be still higher under drought situation. The cost of saving these losses is less than the cost involved in production of same quantity of fruits or vegetables. Hence, it is important to control post-harvest losses by integration of right time of harvesting and adapting appropriate method of harvesting, grading, pre cooling, proper packing, safe transport, development of storage facilities, cooperative marketing system and processing & value addition. In case of arid fruits and vegetables, not many processed products have been commercialised except some aonla products. However, there is good potential of natural dehydration due to availability of ample solar radiation. The practice of sun drying of khejri pods, kair and kachri fruits, etc. is very common in arid Rajasthan. Preservation of fruits and vegetables by Sun drying when they are available in plenty for their use during rest of the year is an age old practice of desert dwellers. This practice needs to be improved and commercialized with application of modern science to get the improved quality products (Fig. 3).

DIVERSIFICATION OF FARMING THROUGH DRYLAND HORTICULTURE

In arid zone more than 90% of the cultivated area is put to arable farming (Bhati, 1994) which yield very low economic returns to the farmers and provide vegetative cover to the soil for a very brief period during the year. The productivity of coarse cereals and other annual crops is uncertain due to low and erratic rainfall and poor soils in dryland. There is substantial shift taking place towards horticulture, agri-horticulture and horti-pasture systems involving fruit trees like ber, pomegranate, custard apple, aonla, guava, tamarind and date palm etc. with annual crops, perennial fodders, short duration fruit crops and vegetables. Introduction of perennial trees and shrubs in existing farming system is desirable for increasing farm income and enhancing drought resilience. Planting of ber at 8mx5m was found ideal for agro horticulture and in such system, grain yield of cluster bean was highest as compared to cow pea and green gram (Anonymous 1994). The yield of both ber and intercrops were higher than monoculture with no adverse effect up to 5 years after planting the ber trees (Singh 1997). Introduction of *Cenchrus ciliaris* in 18 years old ber orchard had no adverse effect on ber with dry fodder yield of 25 q ha^{-1} besides grass seed yield of about 250 kg ha^{-1} and average fruit yield of ber at the tune of 75 q ha^{-1} under rain fed condition. Various multi-storey system with ber, aonla, and khejri have been advocated (Saroj and Yadav, 2001). Growing of Indian aloe (*Aloe barbadensis*) in young ber orchard has also been suggested. The economic analysis of cost and returns of different agro horticulture system revealed highest net profit of Rs. 13,487 ha^{-1} from ber + cluster bean system followed by ber + mung bean (Meghwal and Henry 2008).

Fig. 3. Dehydrated fruits of lasora (left) and khejri (right)

CONCLUSION

The production of dryland horticultural crops largely depend upon the availability and management of limited irrigation and rain water, adopting improved production technology and innovation approaches. Hence, a comprehensive programme is required to develop a viable production system so as to improve yield and quality of produce from dryland horticulture for drought proofing. Making available the planting materials of different horticultural crops suitable for drought prone areas will be a great help. Due consideration need to be given on judicious use of precious and costly available water, *in situ* runoff water harvest and moisture conservation. Using improved biotic and abiotic stress resistant varieties/hybrids, adoption of improved and innovative production techniques including post harvest management technology in a strategic manner can help in minimising the adverse impacts of drought on socio-economic condition and livelihood security of the people.

REFERENCES

Anonymous 1990-91 *Annual Progress Report,* CAZRI, Jodhpur pp 70

Anonymous 1994 *Annual Progress Report,* CAZRI, Jodhpur pp 38-39

Anonymous 1989 *Annual Progress Report,* CAZRI, Jodhpur pp 59

Anonymous 2006-07 *Annual Report*, Central Institute of Arid Horticulture, Bikaner (Rajasthan), pp. 23

Anonymous 2006-07 *Annual Report*, Rajasthan Agriculture University, Bikaner (Rajasthan). pp 74-75

Anonymous 2007 *Annual Progress Report,* CAZRI, Jodhpur pp 87-88

Anonymous 2007-08 *Annual Report*, Rajasthan Agriculture University, Bikaner (Rajasthan)

Awasthi OP, Singh IS and Sharma BD 2006 Effect of mulch on soil-hydrothermal regimes, growth and fruit yield of brinjal under arid condition, *Indian Journal of Horticulture* **63(2)** 192-194

Bankar, G.J. 1989.Vegetative propagation in Annona (*Annona squamosa* L.).*Haryana Journal of Horticultural Sciences* **18(1-2)** 13-14

Bankar GJ and Prasad RN 1989 Bael- a nutritious and useful fruit of arid and semi arid region *Farmers and Parliament* **24(12)** 16-17

Beniwal RK, Soni ML, Yadav ND, Chander P and Talwar HS 2006 Effect of irrigation scheduling on moisture and salt distribution and growth of Kagzi lime under drip irrigation in arid Rajasthan, *Annals of Arid Zone* **45(2)** 169-174

Bhati TK 1994 Agrihorticulture management system. *In: Compedium of lecture of training on Horticultural crops for arid and semi arid areas*, pp 82-90. January 24-31, CAZRI, Jodhpur

Faroda AS 1998 Arid Zone Research: An overview. *In : Fifty years of Arid Zone Research in India*(Eds. A.S. Faroda, and Manjit Singh) pp. 1-16

Gupta JP and Gupta GN 1982 Effect of the application of exfoliated vermiculite on soil properties and vegetable production in arid zone of western Rajasthan, India, *Agrochemica* **26**:38-43.

Gupta JP and Singh SD 1983 Hydrothermal environment of soil and vegetable production with drip and furrow irrigation, *Indian Journal of Agricultural Sciences* **53** 138-142

Meena RS 2008 Effect of GA_3 and NAA on growth, yield and quality of tomato cv. Pusa Ruby grown under semi arid condition, *Current Agriculture* **32(1-2)** 83-86.

Meghwal PR 1997 Exploiting underutilized fruits of arid zones, *Indian Horticulture* **42 (3)** 24-25

Meghwal PR and Vashishtha BB 1998 Effect of time of planting and auxin on rooting of cuttings of ker (*Capparis decidua* L. Forsk), *Annals of Arid Zone* **37 (4)** 401-404

Meghwal PR 2006 Effect of time and intensity of pruning on vegetative growth, fruit yield and quality of phalsa, *Annals of Arid Zone* **45 (1)** 59-62

Meghwal PR, Aseri GK and Rao AV 2006 Raising bio inoculant aided rootstocks for higher yield in ber, *Indian Horticulture* **51(4)** 16-17

Meghwal PR 2007 Propagation studies in lehsua(*Cordia myxa), Indian Journal of Agricultural Science* **77(11)** 765-767

Meghwal PR and Henry A 2008 Economic evaluation of different agro horticulture system under rain fed condition of arid zone. *In: Legumes in dry areas* (Eds. D Kumar, A.Henry and K.P.R. Vittal) Indian Arid Legume Society, Jodhpur and Scientific Publishers Jodhpur, pp 299-303

Reddy NN 1999 National Seminar on. Sustainable Horticulture Production in tribal Regions, CHES, Ranchi, 25-26 July

Saroj PL and Yadav PK 2001 Marushthaliya kshetron ke badalte pariveshmein krishi vaniki ki bhumika(Hindi). *In:Prakartic Sansadhano ke Parbandhan mein Krishi Vaniki ki Bhumika* (Eds. K.S. Dadhwal and K.R Solanki) NRCAF, Jhansi,pp.79-85.

Sharma KD, Pareek OP and Singh HP 1982 Effect of run off concentration on growth and yield of jujube, *Agricultural Water Management* **5 (1)** 73-83

Singh R 1979 *Fruits,* National Book Trust, New Delhi.

Singh SD and Singh Punjab 1978 Value of Drip irrigation compared with conventional irrigation for vegetable products in a hot arid climate, *Agronomy Journal* **70** 945-947.

Singh SD Gupta JP and Singh Panjab 1978 Water economy and saline water use by Drip Irrigation, *Agronomy Journal* **70** 748-751

Singh HP, Singh RP and Singh K 1979 Effect of bentonite sub-surface moisture barrier and run off concentration on soil moisture storage and yield of round gourd grown in sandy soils, *Indian Journal of Agricultural Sciences* **49** 880-885

Singh RS 1997 Note to the effect of inter cropping on growth and yield of ber (*Z. mauritiana* Lank) in semi-arid region, *Current Agriculture* **21(1/2).**

Singh YV, Singh DV and Saxena Anurag 1997 Efficient Management of water and nitrogen by drip fertigation for chilli production in arid region, *Indian Journal of spices* **54** 581-586.

Singh YV, Joshi NL, Singh DV and Saxena Anurag 1999 Response of chilli to water and nitrogen under drip and check basin methods of irrigation, *Annals of arid zone* **38(1)** 9-13

Singh BS and Kumar V 1999 Growth and yield of chilli (*Capsicum annum*) as influenced by application of N and P, *Haryana Journal of Horticulture Science* **28** 112-113

Srivastava BK, Sharma AK, Singh NP and Pandey UB 1984 Effect of organic mulches and irrigation levels on soil temperature, water economy and yield of summer tomato, *Vegetable Science* **11**1-9.

17

Managing Vegetable Crops during Drought

Akath Singh[1] and Pradeep Kumar[2]

INTRODUCTION

Vegetables play a vital role in food and nutritional security of human being and have become an integral part of crop diversification, and in improving economy of farming community. They are important source of valuable protective nutrients especially vitamins, minerals and dietary fibres, which are deficient in many other food items. Besides, they also have good medicinal value of preventive and curative nature. Moreover, rise in income and urbanization, accompanied by affluence and greater health awareness, dietary habits of people shifted towards more intakes of vegetables. According to human dietician about 300 g vegetables including 125 g leafy vegetables, 100 g roots and tubers and 75 g other vegetables per capita per day are required for balanced diet. Though, India ranks second in world's vegetables production with a current production of 162.9 million tonnes from 9.3 million hectare area, is still behind many countries in terms of productivity (17.5 t ha^{-1}) such as China (23.4 t ha^{-1}) and even the current world average (19.4 t ha^{-1}, NHB 2014). In the recent past, concerted efforts have been made to enhance the vegetables production of the country, yet the gap is wider with respect to vegetables requirement and supply. Immense potential therefore, exists in boosting vegetable production even in drought prone areas to increase the national production so as to satisfy present as well as future demands of growing population to have health and nutritional securities, besides improving farmers prosperity. Being succulent in nature as usually consisting of greater than 90% water, vegetables are more sensitive to drought as compared

---

1. Senior Scientist 2. Scientist
Central Arid Zone Research Institute, Jodhpur 342003 (Rajasthan) India
Email: akath2005@yahoo.co.in, pradeephort@gmail.com

to many other crops. Moisture stress, particularly at critical growth period, may drastically reduce productivity and quality of vegetables (Table 1)

MANAGEMENT STRATEGY TO COMBAT DROUGHT

Various management practices have shown the potential to manage vegetable crops under moisture deficit or drought conditions, as follows:

Selection of suitable crops: The foremost step for successful vegetable cultivation in such a fragile ecosystem prone to drought or moisture stress is the selection of crops. Crop selected should have drought tolerant mechanisms like deep root system, tiny foliage to minimize water loss through transpiration, wax coating, hairiness, sunken and covered stomata. The vegetables are grouped on the basis of suitability to different conditions (Table 2).

Table 1. Irrigation needs, critical soil moisture period and impact of water deficit on vegetable crops

Crop	Soil moisture tension (bars)	Available soil moisture (%)	Critical moisture stress period	Major defects caused by water deficit
Onion	-0.25	70	Bulb formation and enlargement	Poor sized bulb and early bolting
Leafy vegetables	-0.30	70	Entire crop duration	Toughness in leaves
Carrot	-0.45	50	Root enlargement	Distorted, rough and misshapen roots
Radish	-0.25	70	Root enlargement	Pithy and distorted roots, accumulation of nitrates
Cabbage	-0.34	60	Head formation and enlargement	Splitting of head and tough leaf
Cauliflower	-0.34	60	Curd formation and enlargement	Ricey curd, buttoning
Tomato	-0.45	50	Flowering and fruit development	Blossom end rot and fruit cracking
Brinjal	-0.45	50	Flowering and fruit development	Blossom end rot, misshapen fruit, poor fruit colour
Okra	-0.70	40	Flowering and pod development	Toughness and fibre development in pods
Chilli and capsicum	-0.45	50	Flowering and fruit development	Low fruit set, shrivelled fruits, blossom end rot
Peas	-0.70	40	Flowering and pod filling	Poor pod filling and shrivelled grains
French bean	-0.45	50	Pod setting and pod enlargement	Poor pod filling with pithy pods
Cucumber	-0.45	50	Flowering and fruit development	Pointed and cracked fruits with bitterness
Water melon	-2.00	40	Flowering and fruit development	Blossom end rot
Sweet potato	-2.00	20	Tuber development and maturity (last month)	Small and Misharpen tuber
Turnip	-0.45%	50	Root enlargement	Woody roots

(Anant Bahadur and Rai 2006)

Table 2. Suitable vegetable crops for specific conditions of arid and semi arid regions

Factors	**Particulars**	**Suitable Crops**
Climatic	Hot & Dry	Melons (muskmelon, watermelon, snap-melon, round melon, *kachri,*), chillies, etc.
Soil texture	Coarse sandy loam	Onion, carrot, radish, sweet potato
Water availability	Rainfed	Cluster bean, snap melon, mateera/ watermelon
	Irrigated	Almost all kinds of tropical/subtropical vegetable
Soil & water quality	Slightly saline	Cole crops, root crops, bulb crops, leafy greens
Nature of cropping system	Fruit based	Beans & melons (rainfed)
		Others (coles, cucurbits, etc.)
	Field crop based	Beans, snap melon (rainfed)

(Doorcenboos and Peruitt, 1992)

Vegetables are further classified into two major groups based on their adaptability and resource based production system under the drought conditions. First those, which are grown as rainfed crop during monsoon season (rainfed vegetables) like mateera/water melon (*Citrullus lanatus*), round melon /tinda (*Praecitrullus fistulosus*), kachri (*Cucumis callosus*), snap melon (*Cucumis melo* var. *momordica*), cluster bean (*Cyamopsis tetragonoloba*) and cowpea (*Vigna unguiculata)*. Other potential vegetable crops which can be exploited under drought conditions are Indian bean (*Lablab purpureus*), sword bean (*Canavalia gladiata*) and ivy gourd (*Coccinia grandis*). Several other indigenous species, which are endemic to the particular region for example in hot arid region khejri (*Prosopis cineraria*), ker (*Capparis decidua*), goonda (*Cordia myxa*), kumat (*Acacia senegal*), guarpatha (*Aloe vera*) may ensure as nutritious vegetables to the people of the region under extremes of the environmental stress.

The other group of the vegetables are adapted under drought by providing assured or limited water for irrigation at least at critical stages and better management practices. The important potential crops are chilli (*Capsicum* sp.), brinjal (*Solanum melongena*), tomato (*Solanum lycopersicum*), okra (*Abelmoschus esculentus)*, onion (*Allium cepa*), garlic (*Allium sativum*), carrot (*Daucus carota*), radish (*Raphanus sativus*), sweet potato (*Ipomea batata*), garden pea (*Pisum sativum* ssp. *hortense*), cole crops like, cauliflower (*Brassica oleracea* var. *botrytis*), cabbage (*Brassica oleracea* var. *capitata*), knol khol (*Brassica oleracea* var. *gongylodes*), leafy vegetables like fenugreek (*Trigonella foenum-graecum*), palak (*Beta vulgaris* var. *benghalensis*) and melons including muskmelon (*Cucumis melo*), kakdi (*Cucumis melo* var. *utilissimus*) and cucumber (*Cucumis sativus*) and gourds including bottle gourd (*Lagenaria siceraria*), ridge gourd (*Luffa acutangula*) and bitter gourd (*Momordica charantia*) and the beans.

SELECTION OF DROUGHT RESISTANT CULTIVARS/ LINES

Different cultivars within a crop species may respond differently to a particular growth environment. There are a few cultivars/lines of vegetables that are identified to have different degree of resistance to drought (Table 3). The Central Institute of Arid Horticulture (CIAH), Bikaner has identified and registered some lines for specific characters such as in cucumber, AHC-13 for small fruit, drought and temperature tolerance, in *Cucumis callosus* (kachri), AHK-119 for high yield and drought tolerance, in snap melon, AHS-10 and AHS-82 for high yield and drought tolerance and in watermelon, AHW 19 for longer shelf-life (Cucurbitaceae 2014). Some of these lines are in commercial cultivation viz., AHK 119 (kachri), AHS 82 (snapmelon) and AHW 19 (watermelon). A drought tolerant tomato line (IIHR-2274) was also identified (Chavan 2007).

Table 3. List of some of the genotypes identified for drought tolerance

Crops/species	Drought tolerant genotypes
Tomato	Arks Vikas, RF-4A
Solanum pennelli	LA 0716, IIHR 14-1, IIHR 146-2, IIHR 383, IIHR 553 , IIHR 555, K-14, Sel-28
Solanum pimpinellifoloium	PI-205009, Pan American, LA1579
Solanum habrochaites	EC520061
S. chilense	LA1958, LA1959, LA1972
Brinjal	Bundelkhand Deshi, SM- 1, SM- 19, SM- 30, Violet Round
Chilli	Arka Lohit, IIHR - Sel.-132
Potato	Alpha, Bintje, Kufri Sheetman, Kufri Sindhuri
Onion	Arka Kalyan, MST 42, MST 46
Cucumber	INGR-98018 (AHC-13)
Snap melon	VRSM- 58, CU 159, CU 196, AHS-10, AHS-82
Kachri	SKY/DR/RS-101, AHK-200
Cucumis melo var. *chat*	Arya
Muskmelon	SC- 15
Sweet potato	VLS6, IGSP 10, IGSP 14, Sree Bhadra

(Modified from Kumar et al 2012)

NURSERY RAISING UNDER PROTECTED CONDITION

The ideal planting materials must have the characters of true to type, resistant/tolerant to biotic and abiotic stress and have the quality of better establishment in field. Green house or low cost polyhouse/low tunnels technology for raising seedling of selected high-value vegetables can circumvent the drought condition and sustain the productivity. It should be sufficiently firm and dense to hold the runners/cutting/seedling. The medium used for nursery beds or bags should preferably have constant volume in both dry and wet conditions, enough moisture retention capacity to avoid frequent watering, porous enough to provide adequate aeration and should have high cation exchange capacity (CEC) for retention of nutrient. Different types of nursery beds may be used for raising seedlings of vegetable crops but during drought or deficit water sunken beds are more preferred for raising seedlings of tomato, brinjal, chilli, cole crops and onion. The seeds must be sown 2-3 cm deep in lines 5-7 cm apart. Before sowing, seeds should be treated with Captan or Thiram @ 2g kg^{-1} of seed. Seedlings prepared in protected conditions must be hardened before transplanting for better setting in the field. Hardening is done by withholding water 4-5 days before uprooting seedlings. Modern nursery raising of vegetable under plug tray in soilless medium containing coco peat, vermiculite and perlite has several advantage to cope the adverse effect of abiotic stress. Healthy seedlings can be raised in short period as compared to the time taken under open field nursery raising. Drastic reduction in the mortality in transplanting of the seedlings was observed as compared to the traditional system of nursery raising. Early planting followed by early harvesting is possible to escape the drought period for raising such nursery under protected condition.

GRAFTING AS A TOOL TO IMPROVE DROUGHT TOLERANCE

Continued successful management of the limited amount of water available for agricultural uses, in particular vegetable production depends upon the better agronomic practices and enhanced understandings of water productivity (Karam et al 2011). Moisture stress depresses crops yield due to decreased nutrient uptake by the root and transport to the shoot because of a restricted transpiration rate affecting active transport and membrane permeability (Kramer and Boyer 1995). One way to reduce losses in production and to improve water productivity under drought conditions in high-yielding vegetable genotypes would be grafting them onto rootstocks capable of reducing the effect of water stress on the shoot as observed in tree crops (Garcia-Sanchez et al 2007). Grafting is an age old practice and has long been most commonly applied in many woody plant species especially, in fruit crops. In vegetables, it is relatively new tool and has commercially been employed for few decades only with primary aim of developing plant tolerance to soil borne diseases. Now, with the widening of objectives of increasing plant tolerance to other biotic and abiotic stresses, it has become popular and spread all over the world and being commercially employed in fruiting vegetables belonging to cucrbitaceae (cucumber, watermelon and muskmelon) and solanaceae family (tomato, pepper and eggplant) and also in beans. Nowadays, grafting is used to enhance the tolerance against abiotic stresses (Colla et al 2010). Sanders and Markhart (1992) have shown in beans (*Phaseolus vulgaris* L.) that the osmotic potential of dehydrated scions of grafted bean plants was determined by the rootstocks, while the osmotic potential of non-stressed scions was governed by the shoot. Drought tolerance provided by either the rootstock or the scion resulted in enhanced nitrogen fixation in soybean (*Glycine max* L.) (Serraj and Sinclair 2002). Grafting tomato (cv. BHN 602) onto rootstock 'Kkung Jjak' increased qualitative drought-tolerance as compared to non-grafted tomato plants (Nilsen et al 2014). Grafted mini-watermelons onto a commercial rootstock 'PS 1313'_ (*Cucurbita maxima* Duchesne × *Cucurbita moschata* Duchesne) revealed more than 60% higher marketable yield when grown under conditions of deficit irrigation compared with ungrafted melons (Rouphael et al 2010). The higher marketable yield recorded with grafting was mainly due to an improvement in water and nutrient uptake, (N, K, and Mg), and higher CO_2 assimilation. Grafting experiments with ABA deficient mutants of tomato showed that stomata can close independently of the leaf water status suggesting that there is a chemical signal produced by the roots that controls stomatal conductance (Holbrook et al 2002). Combined use of genetic and agronomic technologies to sustain water status may maintain plant growth under conditions where yield would otherwise be significantly reduced.

SOIL MOISTURE CONSERVATION

Various management practices help to conserve soil moisture *in-situ* enabling sustained vegetable production even during drought. The use of organic and inorganic mulches is common in high-value vegetable production systems. Mulching helps in maintaining soil temperature and moisture, avoiding weed competition, improving soil structures and biological activities, resulting into increased crop yield. These protective coverings help reduce evaporation, moderate soil temperature, reduce soil runoff and erosion, protect vegetables from direct contact with soil and minimizes weed growth. In addition, the use of organic materials as mulch can help enhance soil fertility, structure and other soil properties. When compared to other mulches plastic mulches are completely impermeable to water; it therefore prevents direct evaporation of moisture from the soil and thus limits the water losses. In this manner, it plays a positive role in water conservation. Under arid conditions it is reported that, organic mulches like grasses, weeds and crop residues reduce the soil temperature considerably during summer months, while during winter months clear/ transparent plastic mulch increase soil temperature. Black coloured plastic mulch controls weed growth while silver coloured plastic reflects lights, act as effective repellent for insects, birds, etc.

The rainfall is conserved *in-situ* by adopting suitable land use and appropriate conservation measures, which increase infiltration rate in soil and cut down runoff flowing over a slope. Though there are many *in-situ* water conservation techniques, the more effective in arid climate are : ridges and furrow system, graded furrows, broad bed and Furrow system, sunken and raised bed system of planting.

WATER-SAVING IRRIGATION

The quality and efficiency of water management determine the yield and quality of vegetable products. Water demand is a function of weather conditions, crop species and variety, stage of growth, soil water retention capacity and texture and irrigation system management. If water is scarce and supplies are erratic, then irrigation at critical stage and soil moisture conservation are the most important agronomic interventions to maintain yields during drought stress. Among the different systems, drip irrigation is an efficient tool of water application in most of vegetables, particularly during drought period. In this system water is delivered through drippers directly to the soil adjoining to the root system, which absorb the water immediately. Studies indicate that drip irrigation could save water from 33 to 55% with yield increase to a tune of 50% or more besides improving the quality of the produce in different vegetable crops (Table 3). Fertilizers and pesticides can also be delivered with drip irrigation system to the crops and thus minimizes their application losses. Most importantly, their split application can be made as per the requirement at different crop stages with crucial input during drought period.

Results of a recent study conducted at CAZRI, Jodhpur revealed that under limited water supply (0.6 ETc), the yield depression in chilli crop and cluster bean in a chilli-clusterbean cropping system could be significantly minimized with the substitution of 50% N as organic manures through vermicompost along with 50% NPK as fertilizers . In another study, N substitution through manures (compost or vermicompost) at 25%-75% along with RDF enhanced carrot yield and quality parameters under traditional irrigation supply conditions, besides improving soil properties (Kumar et al 2014). Nonetheless, crops should be irrigated in the evening to save evaporation losses and as a matter of fact most vegetables do most of their growing at night.

Table. 4. Yield increase and water savings under drip irrigation over conventional irrigation

Crop	Yield increase (%)	Water savings (%)
Brinjal	14	53
Cabbage	2	60
Beet root	7	79
Okra	16	40
Ridge gourd	17	59
Bitter gourd	39	53
Chilli & capsicum	44	62
Radish	2	77
Tomato	50	39
Water melon	88	46

(Anant Bahadur and Rai 2006)

NUTRIENT MANAGEMENT

Soils of arid region are poor in organic matter that affects nutrient use efficiency as well as water retention. Vegetables are nutrient exhaustive crops and deplete soil fertility extensively thus necessitating the judicious application of fertilizers. Organic fertilization can improve vegetable crop tolerance to soil stresses like salinity, drought, alkalinity and excessive amounts of toxic elements. A number of studies have shown that organic manure application increases water as well as nutrient use efficiency due to increased organic matter content and other soil physical, chemical and biological properties which in turn favour better plant growth and development that help in counteracting the adverse conditions in the root environment. Under drought conditions, organically fertilized crops produce higher yields than comparable crops managed conventionally and among several underlying mechanisms, higher water holding capacities in organically managed crop (Liebig and Doran 1999, Wells et al 2000) is believed to be major one.

Foliar application of nutrients has also been recommended as a means of improving plant growth especially in the early stages of development as it provides the required nutrient directly to the location of demand in the leaves and results in rapid absorption, being independent of root activity and soil water availability. Beneficial effects of foliar application of mineral nutrients on vegetable crops have been studied under several stress conditions in the root zone (e.g. drought, salinity and alkalinity). Organic manure application improves soil physical properties as evident from improved soil structure and reduced bulk density and penetration resistance (Mishra and Saha 2008). It also increases the N-use efficiency of the applied fertilizers (Aggarwal and Praveen Kumar 1993). Vermicompost gained popularity due to recent interest in organic farming and is mostly used in fruits and vegetables (Prasad 2008). It is a rich source of both micro and macronutrients, vitamins, growth hormones and enzymes (Bhavalkar 1997). Organic content was found to be significantly higher with the application of FYM and vermicompost, while bulk density decreased. In arid regions, where soils are poor in organic matter content, moderating organic matter favourably affects nutrient use efficiency, water holding capacity and ultimately improves crop yield.

MYCORRHIZAL FUNGI INOCULATION

One way to avoid or reduce losses in production caused by different abiotic stresses in high-yielding vegetable genotypes could be to inoculate them with microbial inoculants that are capable of reducing the detrimental effect of external rooting medium on crop performance. Arbuscular mycorrhizal (AM) fungi are soil microorganisms that establish symbiotic associations with fine young roots of the majority of higher plants, providing a direct physical link between soil and plant roots. AM fungi have been shown to decrease yield losses of plants in drought conditions. This may be due to increased uptake of nutrients with low mobility such as P, Fe, Cu & Zn and decrease the uptake of Na (Al-

Karaki 2006) in AM plants. In addition, AM fungi can improve physiological processes like water absorption capacity of plants by increasing root hydraulic conductivity and favourably adjusting the osmotic balance and composition of carbohydrates and by optimizing the hormonal balance (Rosendahl and Rosendahl 1991, Ruiz-Lozano 2003). Moreover, it is likely that the contribution of the AM symbiosis to plant drought tolerance results from a combination of physical, nutritional, physiological and cellular effects. Although different plant species can vary in their sensitivity and responses to the decreased water potential caused by water deficit, it is assumed that all plants have an encoded capability for stress perception, signalling and response. Subramanian et al (2006) reported that AM colonization improved drought resistance of field grown tomato plants as a consequence of enhanced nutritional status especially P and N.

USE OF PLANT GROWTH REGULATORS, ANTITRANSPIRANTS AND OSMOPROTECTANTS

Phytohormones are chemical messengers produced in one part of plant and translocated to the other parts, where they play critical roles at extremely low concentration in regulating plant responses to stress. These are natural products and when synthesized chemically, they are called plant growth regulators (PGRs). Exogenous application of PGRs, helps in alleviating the adverse effects of various abiotic stress and also improves germination, growth, development and seed yields and yield quality (Javid et al 2011). Application of some of the PGRs will prove beneficial for better crop growth and development when grown under water deficit situations. Some of the PGRs and their effects on crops in order to suit to the water stress conditions are:

- Cycocel and Mepiquat chloride: Promote root growth (for more water absorption), suppress leaf area development (for reducing transpiration loss of water) and delay the onset of leaf senescence.
- Cytokinins and Salicylic acid: Delay the leaf senescence processes and also favour stem reserve utilization by developing grains especially during the water deficit situations.
- Brassinolides: They increase the photosynthetic activity of the plants.
- Ascorbic acid: Acts as an anti-oxidant for scavenging reactive oxygen species (ROS) accumulating under stress and thus avoiding membrane damage.
- Abcisic acid (ABA) : They are also called dormins and act as anti-gibberellins. It is synthesized in leaves of many plants and responsible for closing stomata during drought conditions, hence act as plant stress harmone.

Prolonged drought can drastically reduce the yield in different vegetables. But, intermittent drought for 10 – 15 days at early or late stage is common under rainfed conditions. Drought during the critical phenological phase like flowering and fruit development is highly detrimental. However, the crop productivity is dependent on how fast a plant can recover after a stress of 6-10 days. The severity of intermittent drought of 6-10 days during critical stages of the crop can reasonably be avoided by the use of antitranspirants and thus crops can be saved. Antitranspirants (kaolin, paraffin, calcium bicarbonate, lime water) can effectively be used in the crops under water stress due to deficit rainfall.

Plants have evolved various protective mechanisms that allow them to acclimatize to unfavourable environments for continued survival and growth. One such mechanism that is ubiquitous in plants is the accumulation of certain organic metabolites of low molecular weight that are known collectively as compatible solutes or osmoprotectants. These compounds include proline, ectoine, betaines, polyols, and ehalose, and have evolved in many different organisms. Since some crop plants have low levels of these osmoprotectants or none at all, engineering osmoprotectant biosynthesis pathways is a potential way to improve stress tolerance. Exogenous application of potential osmoprotectants such as proline is an important approach to alleviate adverse effects of abiotic stresses on plants and it has gained considerable attention in mitigating the effect of water deficit (Ali et al 2007). Proline may have profound effects in ameliorating the adverse effects of water deficit on growth of vegetables, as reported in maize, where it had enhanced the photosynthetic rate of water stressed plants, which was positively associated with sub-stomatal CO_2 and stomatal conductance as well as photosynthetic pigments (Ali et al 2007). Furthermore, it can effectively regulate osmotic potential and plays a vital role in sustaining plant growth under osmotic stress (Serraj and Sinclair 2003, Ali et al 2007, Ashraf and Foolad 2007).

CONCLUSION

Due to limited availability of arable land and the increasing market demand for vegetables around the world, vegetables are compelled to be cultivated in unfavourable conditions including arid and semi-arid conditions, where drought or moisture stress are common feature that diminish production and quality of vegetables. However, with the adoption of appropriate techniques and following integrative approaches of vegetable production, such as selection of suitable crop and variety or use of seedlings grafted onto suitable drought hardy rootstocks, application of proper nutrition with the use of bio-inoculant (e.g., mycorrhiza) and plant growth regulators, antitranspirants, and the crop production with drip and mulches, the adverse effects of drought could be minimized to a greater extent.

REFERENCES

Aggarwal RK and Praveen Kumar 1996 Integrated use of farmyard manure and fertilizer nitrogen for sustainable yield of pearl millet in a desert sandy soil, *Annals of Arid Zone* **35** 29-35

Ali Q, Ashraf M and Athar HR 2007 exogenously applied proline at different growth stages enhances growth of two maize cultivars grown under water deficit conditions, *Pakistan Journal of Botany* **39(4)** 1133-44

Al-Karaki GN 2006 Nursery inoculation of tomato with arbuscular mycorrhizal fungi and subsequent performance under irrigation with saline water, *Scientia Horticulturae* **109** 1–7

Anant Bahadur and Rai M 2006 Use drip irrigation for higher yield in vegetables, *Indian Horticulture*, **51(3)** 8-9

Ashraf M and Foolad MR 2007 Roles of glycine betaine and proline in improving plant abiotic stress tolerance, Environmental and Experimental Botany **59** 206-16

Bhavabakar U 1997 Vermiculture technology for LEISA Seminar on low external input sustainable agriculture, Amsterdam, Netherland.

Chavan ML 2007 Drought tolerance studies in tomato (*Lycopersicon esculentum* Mill.), Department of Crop Physiolgoy College of Agriculture, Dharwad University of Agricultural Sciences, Dharwad. 133-135 pp

Colla G, Rouphael Y, Leonardi C and Bie Z.2010 Role of grafting in vegetable crops grown under saline conditions, *Scientia Horticulturae* **127** 147-55.

Garcia-Sanchez F, Syvertsen JP, Gimeno V, Botia P and Perez-Perez JG 2007 Responses to flooding and drought stress by two citrus rootstock seedlings with different water-use efficiency, *Physiol Plant* **130** 532-42

Holbrook NM, Shashidhar VR, James RA and Munns R 2002 Stomata control in tomato with ABA deficient roots: response of grafted plants to soil drying, *Journal of Experimental Botany* **53** 1503-14

Karam F, Saliba R, Skaf S, Breidy J, Rouphael Y and Balendonck J 2011 Yield and water use of eggplants (*Solanum melongena* L.) under full and deficit irrigation regimes, *Agricultural Water Management* **98** 1307-16

Kramer PJ and Boyer JS 1995 *Water relations of plants and soils.* Academic Press, San Diego, pp 495–524

Kumar P, Meghwal PR and Painuli DK 2014 Effect of organic and inorganic nutrient sources on soil health and quality of carrot, *Indian Journal of Horticulture* **71(2)** 222-226

Kumar R, Slankey SS and Singh M 2012 Breeding for drought tolerance in vegetables, *Vegetable Science* **39(1)** 1-15.

Liebig MA and Doran JW 1999 Impact of organic production practices on soil quality indicators, Journal *of. Environmental Quality* **28(5)** 1601-09

Javid MG, Sorooshzadeh A, Moradi F, Mohammad SA, Sanavy M and Allahdadi I 2011 The role of phytohormones in alleviating salt stress in crop plants, *Australian Journal of Crop Science* 726-734

Mishra VK and Saha R 2008 Soil physical behaviour and rice (*Oryza sativa*) yield under different sources of organics, methods of puddling and zero tillage, *Indian Journal of Agricultural Sciences* **78** 399-404

NHB 2014 Horticulture database 2014. National Horticulture Board. Gurgaon, India.

Nilsen ET, Freeman J, Grene R and Tokuhisa J 2014 A rootstock provides water conservation for a grafted commercial tomato (*Solanum lycopersicum* L.) line in response to mild-drought conditions: A focus on vegetative growth and photosynthetic parameters. *PLoS ONE* 9(12): e115380. doi:10.1371/journal.pone.0115380.

Prasad R 2008 Integrated plant Nutrient supply system (IPNS) for sustainable agriculture, *Indian Journal of fertilizers* **4(12)** 71-90.

Rosendahl CN and Rosendahl S 1991 Influence of vesicular arbuscular mycorrhizal fungi (*Glomus* sp.) on the response of cucumber (*Cucumis sativus*) to salt stress, *Environmental and Experimental Botany* **31** 313–18.

Rouphael Y, Schwarz D, Krumbein A and Colla G 2010 Impact of grafting on product quality of fruit vegetables, *Scientia Horticulturae* **127** 172–79

Ruiz-Lozano JM 2003 Arbuscular mycorrhizal symbiosis and alleviation of osmotic stress. New perspectives for molecular studies. *Mycorrhiza*, 13 online first published on April 11, 2003; ODI: 10. 1007/s00572-0030237-6.

Sanders PL, Markhart AH 1992 Interspecific grafts demonstrate root system control of leaf water status in water stressed *Phaseolus, Journal of Experimental Botany* **43** 1563–1567

Serraj R and Sinclair TR 2002 Osomlyte accumulation: can it really help increase crop yield under drought conditions, *Plant Cell Environ* **25** 333-41

Subramanian KS, Santhanakrishnan P and Balasubramanian P 2006 Responses of field grown tomato plants to arbuscular mycorrhizal fungal colonization under varying intensities of drought stress, *Scientia Horticulturae* **107** 245–53

Wells AT, Chan KY and Cornish PS 2000 Comparison of conventional and alternative vegetable farming systems on the properties of a yellow earth in New South Wales, *Agriculture Ecosystem and Environment,* **80(1-2)** 47-60.

18

Agroforestry and Integrated Farming System

S.P.S. Tanwar and J.C. Tewari

INTRODUCTION

Drought is experienced *albeit* varyingly, in almost all parts of the earth. It can be defined as the state of adverse and widespread hydrological, environmental, social and economic impacts due to less than generally anticipated rainfall. Over 68% of India is vulnerable to drought, of which 33% (rainfall< 750 mm) is designated 'chronically drought prone' while another 35% (rainfall 750-1125 mm.) is 'drought prone'. The frequency of drought is maximum in arid regions. Out of 50 years (1960-2009), the drought frequency varied from 32 years at Barmer to 24 years at Hanumangarh (Rao et al 2012). Broadly people in arid regions of India experience drought in two to three years out of five and it continues to take a heavy toll of region's economy as well as human and livestock population.

Depending on its severity, drought causes different degrees of suffering to the human societies, especially those depending more on agriculture. For instance, the drought of 2002 was disastrous when the annual rainfall was a mere 127 mm (38% of long term average), and the crucial summer rainfall was 90 mm, against normal of 280 mm. Vulnerability of a household unit or of a community can be defined as its capacity to withstand the exposure to drought stress, either experienced or anticipated (Saha et al 2012). Since food, fodder, water and livelihoods are the major casualties of the drought; ensuring their availability is a prerequisite to reduce vulnerability.

Annual crops are most affected by quantum and distribution of rainfall. To minimize the risk and uncertainty, farmers in arid regions practice farming systems in one form or other. In fact traditional farming systems represent a crop

Principal Scientists
Central Arid Zone Research Institute, Jodhpur 342003 (Rajasthan) India
Email: spstanwar@gmail.com

– livestock - tree/shrubs – human continuum. The areas falling in <250 mm rainfall zone have predominance of grasses and shrubs; hence range/ pasture development with livestock rearing is the major proposition (Bhati and Joshi 2007). In areas receiving rainfall between 250-350 mm besides grasses and shrubs, multipurpose tree species dominate and mixed farming encompassing agroforestry system, mixed cropping, livestock and pasture management are main livelihood options. In areas receiving rainfall >300 mm, crops and cropping system diversification, agroforestry and livestock rearing are major systems of sustenance of arid zone farmers (Fig. 1). These local survival systems some 80 years back were sustainable but now become inadequate due to technological advances and continuously increasing requirements of household.

Agroforestry, the integration of trees and shrubs with annual crop production, is an age old system of arid agriculture. It provide shade, a steady supply of food and/or income throughout the year, arrest degradation and maintain soil fertility, diversify income sources and stabilize it, enhance efficiency of soil nutrients , water & radiation and provide regular employment.

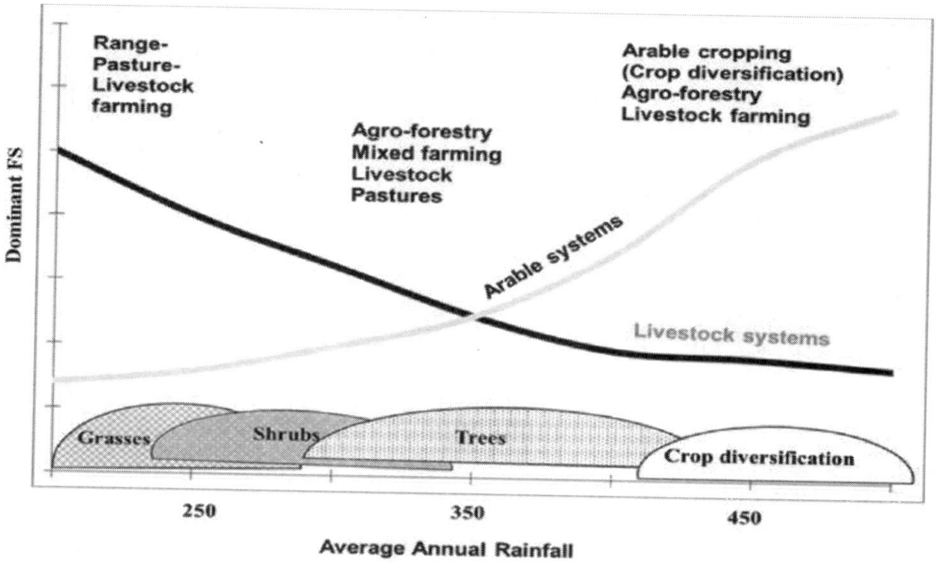

Fig. 1. Dominance of various farming systems according to rainfall pattern in arid zone

TRADITIONAL AGROFORESTRY SYSTEMS

Traditionally, large tracts of lands in arid and semi-arid regions of Rajasthan have widely scattered trees of various species and in association with them crops are grown. Agroforestry ensures top feed resources and economic returns from fruits, fuel wood, timber etc. during the drought years. Besides, trees like *Prosopis cineraria* (*Khejri*) are unique in their ability to improve crop yields grown beneath their canopy. Desert dwellers have evolved agroforestry based drought protective mechanisms through their ingenuity and centuries old accumulated knowledge. *Prosopis cineraria* based agroforestry system is most popular and widespread covering 60% area of arid zone. Rainfall appears to be the governing factor for evolution of traditional agroforestry systems. Four types of agroforestry systems have been identified in arid regions. Over 400 mm rainfall zone, the predominant system is *P. cineraria – Acacia nilotica* based. In 200-300 mm rainfall zone *Ziziphus* spp. – *P. cineraria* based system is widely prevalent where *Z. numularia* is the main species with few trees of *P. cineraria* and *Ziziphus rotundifolia*. Shrubs dominate < 200mm rainfall zone mainly *Ziziphus* sp. (>65%) along with scattered trees of *Prosopis cineraria* and *Salvadora sp.* (Narain and Tewari 2005, Tewari et al 2014). With the decrease in rainfall, the density of woody component of the system decreased substantially. *Tecomella undulata, Hardwickia binnata, Ailanthus excelsa* and *Holoptelea integrifolia* are the other important tree components that may find place in IFS.

TREE CROP INTERACTIONS IN AGROFORESTRY SYSTEMS

Moderation of microclimate

Agroforestry finds prominent place in arid zone due to unique climatic-tree- crop interactions. Trees play a very important role in moderating the microclimate and soil fertility in arid regions. They do so by influencing radiation flux, air temperature, wind speed and saturation deficit of under storey crops. All of these changes have a significant impact on modifying the rate and duration of photosynthesis, subsequent plant growth, transpiration and soil water use (Monteith et al 1991). Following examples present microclimatic moderation effect in harsh climatic conditions of arid regions of Rajasthan.

Air temperature exhibited appreciable variation under the canopy of *Acacia tortilis* during monsoon period (Tewari et al 1989). A decline of 0.1°C to 0.7°C was evident beneath the canopy cover of *Acacia tortilis* than that of surrounding open areas at 07 hrs. Around 14 hrs, this decline was of the order of 0.6°C to 2.0°C. Similarly, it was found during monsoon season that the soil temperature just beneath the tree cover was lower by as much as 10°C to 16°C in top soil zone and 4°C to 5°C at 30 cm depth when compared to open field conditions, thereby indicating better soil-thermal regime. Seven year old *Acacia tortilis* trees provided more ambient humidity to cluster bean grown beneath it. This finding proves the efficiency of agroforestry practices for maintaining favourable moisture status on arable lands (Ramakrishna and Shastri 1977).

Enhancement in soil fertility

Soil fertility under tree species (*Prosopis cineraria* and *Acacia nilotica*) and in open field conditions was studied intensively (Singh and Lal 1969). The organic carbon content was higher under the canopy of both the species. The available N, P & K were maximum beneath *P. cineraria* followed by *Acacia nilotica.* Burman et al (2002) investigated effect of fourteen tree species prevalent in arid zone and concluded that *P. cineraria* and *Hardwickia binata* which lead to higher built up of amino acids, amino sugars, hydrolysable NH_4 –N and total hydrolysable N are likely to be more beneficial to the companion crop than rest of the species studied.

Soil and moisture conservation

In arid regions, higher temperatures and drier conditions lead to lower organic matter accumulation in the soil resulting in poor soil structure, reduction in infiltration of rain water and increase in runoff and erosion (Rao et al 1998) while the expected increase in the occurrence of extreme rainfall events will adversely impact on the severity, frequency, and extent of erosion (WMO 2005). These changes will further exacerbate an already serious problem is being faced by arid regions of the country.

The woody component in agroforestry systems helps in reducing soil erosion which is the most harmful abiotic stress in arid region. It was observed if rows of trees are planted right angle to wind direction a tremendous amount of soil drop can be checked from agricultural fields. The shelterbelts on agricultural fields form a type of agroforestry practice (Gupta et al 1984). Data mentioned in Table 1 shows the effectiveness of agroforestry in controlling the soil erosion in arid regions.

Table 1. Effect of different type of shelterbelts on soil erosion

Type of shelterbelt	Total amount of soil loss (kg ha^{-1})		
	Year I	**Year II**	**Mean**
Prosopis juliflora	93.2	609.3	351.2
Cassia siamea	91.5	277.1	184.2
Acacia tortilis	106.0	494.1	300.0
Agricultural field without trees	262.7	831.0	546.8

There are several mechanisms whereby agroforestry may use available water more effectively than the annual crops. Firstly, unlike in annual systems where the land lies bare for extended periods, agroforestry systems with a perennial tree component can make use of the water remaining in the soil after harvest and the rainfall received outside the crop season. Secondly, agro-forests increase the productivity of rain water by capturing a larger proportion of the annual rainfall by reducing the runoff and by using the water stored in deep layers. Thirdly, as discussed earlier, the changes in microclimate (lower air temperature, wind speed and saturation deficit of crops) reduce the evaporative demand and make more water available for transpiration.

The tree canopies in agroforestry systems intercept the rain and reduce runoff (Khan et al 1995). In a study at CAZRI, it was found that in *Acacia tortilis* silvi-pasture stand canopy interception was 21.4%, whereas in *Colophospermum mopane* silvi-pasture it was 13.1%. Rainfall interception was positively related with canopy cover and negatively related with throughfall. Average surface run off in *A. tortilis* based silvi-pasture stand was 53% higher than in *C. mopane* based silvi-pasture. This indicated that hydraulic response to rain is dominated by plant species character however; the percent annual runoff and soil erosion was very low in situations with trees on agricultural fields in comparison to bare soil condition. Thus, the enhanced use efficiency of rain water by woody species in agroforestry systems improves agricultural productivity.

ENHANCING LIVELIHOOD OPTIONS THROUGH AGROFORESTRY

Agroforestry contributes to livelihood improvement in arid regions of India, where people have a long history of accumulated local knowledge. Arid regions are particularly notable for ethano-forestry practices and indigenous knowledge system on growing trees on farm lands. The economics of previously discussed traditional agroforestry systems of arid western Rajasthan indicated that net B:C ratio of such systems is generally on positive sides (Table 2).

Table 2. Economics of traditional agroforestry systems of arid western Rajasthan

Agroforestry system	Expen-diture (Rs/ha)	Returns (Rs/ha)			Gross returns (Rs/ha)	Net Returns (Rs/ha)	Net B:C ratio
		Crops	Fuel wood	Leaf fodder			
P. cineraria-A. nilotica based	1850	4103	1230	870	6203	4353	2.3
P. cineraria based	1550	3670	600	420	4690	3140	2.0
Z. spp. - P. cineraria based	1550	1506	620	600	2726	1176	0.7
Z. spp. - P. cineraria - Salvadora spp. based	1500	1400	500	500	2400	900	0.6

In fact, multipurpose trees in agroforestry systems provide a rational formula for drylands. In hot arid regions trees on farmlands are the organisms with an ecological investment strategy (Oldeman 1983). Matter and energy are immobilised in long lived trees and this majority of accumulated production is harvested at longer interval than in case of arable crops. Pearlmillet and cowpea perform better with *khejri* than greengram and dewgram. Besides good yield of dryland crops, bonus yield of dry leaves and twigs (650-1050 kg ha^{-1}) and fuel wood (1.8-2.6 t ha^{-1}) could be obtained from tree through annual lopping (Bhati et al 2008). A tree density of 100-200 plants ha^{-1} was found optimum for minimum interference with yield of dryland crop under *P. cineraria* canopy shade.

The most spectacular evidence in context of livelihood improvement potential of agroforestry in arid region came through *Acacia senegal* based localized traditional agroforestry system in parts of Barmer and Jodhpur districts (15 villages). *Acacia senegal* is source of gum arabic which has very high commercial value. The farmers harvest substantial quantity of gum arabic through a small intervention by CAZRI ie. Gum induction using ethephon injection. CAZRI gum inducer has been so popularised in the villages that farmers have made their own cooperatives for purchasing of gum inducer and sale of produce in the market. During the year 2008-09, 12,000 trees were treated through CAZRI gum inducer which resulted in exudation of 5.4 tonnes of gum Arabic and farmers earned Rs. 27 Lakhs from the sale of gum. Subsequently in the year 2009-10 and 2010-11, 21,000 and 22,600 trees were treated, respectively resulting in production of 10.5 t and 7.6 t gum respectively. In the process farmers earned 52 Lakhs in the years 2009-10 and Rs. 38 Lakhs in the year 2010-11. In this way the farmers of 15 villages earned Rs. 117 Lakhs from the sale of gum Arabic with the average sale price in local market Rs. 500/kg.

Jujube (*Ber*) is another very useful agroforestry species in arid regions. The leaves make excellent fodder, crude protein (13-17%) and fiber (15%). The branches and twigs are excellent fuelwood with high calorific value (4878 kcal kg^{-1}). The fruits are delicious and rich source of vitamin A, C and B complex. Introduction of *Zizypus mauritiana* var. Gola & Seb has transformed horticulture scenario of western Rajasthan and now maximum orchards are of these two varieties fetching more than a lakh rupees per hectare as net returns. Ber can be fitted in many systems i.e. sole plantation, *ber* + crops as well as *ber* + grasses. *Ber* based pasture system can be a major source for livelihood in many parts of arid zone. This system provides production in terms of human food, livestock feed, and fuel wood needs of communities.

These examples demonstrate the potential of agroforestry in reducing vulnerability in arid regions. In addition to providing food and feed, there are numerous other non-timber tree products (excluding gums and resins) collected from trees on farm lands. *Prosopis cineraria* trees in addition to providing fuel and fodder give nutritious vegetable in form of pod, locally known as '*sangri*'. Likewise seeds of *Acacia senegal* are also used in preparing vegetable. During drought years their value is increased tremendously. In fact, scattered trees on agricultural fields in arid regions of western Rajasthan are multipurpose in nature and they have ability to satisfy the expectations and aspiration of rural folk regarding production of their basic needs. Thus, agroforestry provides unique basis of life support and is an integral part of farming systems of arid regions.

INTEGRATED FARMING SYSTEM

No single enterprise can impart stability in income in fragile environments. Adopting a suitable mix of enterprises in farming is the first step towards preparedness to drought. The concept of sustainable farming system is holistic and far sighted. First it aims to optimize long term farm output to develop a range of flexible contingency plans from which to select those best suited to the rainfall in a given season, be it high or low. Second principle is the optimal management of water resources be they soil moisture, ground water or stored water for maximum long term stability along with minimal short term fluctuations. Finally, the farming system must be resilient, that is able to endure drought and speedy recovery from it.

Integrated farming system (IFS) approach has been widely advocated as a means to harmonize use of scarce resources so as to make the production system sustainable and drought resilient. IFS besides increasing system productivity also envisages harnessing complementarities and synergies among different agricultural sub systems/enterprises and augmenting the total productivity, profitability, sustainability and gainful employment for a house hold. Meeting the requirement of food, fodder and livelihood at an optimal level during adversities like drought and exploiting full potential of favourable years is at the core of the concept of IFS. It works on strategy of diversification of system and crops. To impart resilience in arid zone farming, perennial components are included in the system and land is put to alternate use as per capability and requirements. This includes trees, shrubs and grasses along with annual crops. Agroforestry has been discussed in detail earlier, other components of integrated farming system in arid zone *vis a vis* drought resilience potential are discussed below.

CROP DIVERSIFICATION

Diversification of crops and cropping systems seems to be a viable option to impart resilience in agricultural production system. To cushion the adverse effects of drought and uncertainty, mixed sowing of seeds of various dryland crops viz., pearlmillet (*Pennisetum typhoides*), clusterbean (*Cyamopsis tetragonoloba*), mothbean (*Vigna aconitifolia)* and sesame (*Sesamum indicum*) in an approximate ratio of 8:2:2:1 (by weight) is a common practice (Bhati and Faroda 2000). If drought commences, they cut down pearlmillet at young stage even to be fed to the animal. However, this practice of mixed cropping hampers efficient management of crops from weeding to harvesting. During timely onset of monsoon, farmers usually take pearlmillet on more than 60% area and if monsoon delays (beyond July 15) *kharif* pulses (*Vigna radiata*, *Vigna aconitifolia*) get preference. Based on climatic data, vulnerability to drought and changing food habits, it is advocated to reduce area under pearlmillet to 40% and allocate rest of the area to *kharif* legumes (30%), oilseeds (15%) and forage crops (15%).The crop diversification should be taken both as sole crop and intercropping systems. Inclusion of more efficient, competitive and high value-low volume commodities like medicinal, dye yielding crops could provide much buoyancy in arid farming. This includes *Cassia angustifolia, Aloe vera, Plantago ovata* and *Lawsonia* (henna). Cultivation of *Aloe* in strip plantation also arrested soil erosion. Aloe resulted in net soil deposition of 179.2 t ha^{-1} against soil loss of 248.3t ha^{-1} from bare soil, at Bikaner (Rathore et al 2008).

AGRI –HORTICULTURE

Horticulture based production system is considered effective strategy for improving productivity, employment opportunities, economic condition and nutritional security. Several drought hardy fruit trees/shrubs like *Ziziphus mauritiana, Ziziphus nummularia* var. *rotundifolia, Cordia dichotoma, Capparis decidua, Salvadora oleoides,* are suitable for the area receiving rainfall <300 mm. Several other fruits such as *Emblica officinialis, Punica granatum, Aegle marmelos, Phoenix dactylifera* and *Tamarindus indica* can be grown in the area having irrigation facilities. Among the vegetable crops *Solanum melongena, Lagnareria siceraria, Luffa acutangula, Luffa cylindrica, Citrullus lanatus, C. lanatus var. fistulosus, Cucumis melo* var. *utillismus, Cucumis melo* var. *momardica, Cucumis callosus, Moringa oleifera, Cyamopsis tetragonoloba* and *Vigna unguiculata* are suitable for horticultural based framing systems (Pareek and Awasthi 2008). Agri-horti system comprising *Ziziphus* + greengram was found very promising in <250-300 mm rainfall zone (Faroda 1998, Gupta et al 2000, Sharma and Gupta 2001, Singh et al 2003). Intercropping of bottle guard during *kharif* season and pea (arkel) and kasuri methi during *rabi* with *ber* plantation did not cause adverse effect on three year old ber (Singh and Kumar 1993). A study on kinnow based agri-horti system under irrigated arid condition at Sri Ganganagar showed that intercrop did not show any significant negative effect on fruit yield (Bhatnagar et al 2007). Pomegranate has been found compatible with *pearl millet*, greengram, Isabgol, sorghum and cumin in Jalore district of Rajasthan (Gupta, 2000).

AGRI-PASTURE

Cropping between grass strips laid out against prevalent wind direction not only reduce soil erosion but also provide at least some biomass during low rainfall years. In good rainfall year, production of arable crops increase along with increased forage yield of grasses. This practice of strip cropping of legumes and grasses holds promises for Bikaner region (Singh 1989, 1995). A strip cropping of grasses and *kharif* legumes in 1:2 ratio has been recommended, with a strip width of 5 m. Ley farming also increased grain yield of crop significantly over control (Singh and Gupta 1997). Intercropping of arid legumes with *Cenchrus ciliaris* gave higher yield, return and moisture use efficiency than sole pasture (Dauley 1994).

SILVI-PASTORAL SYSTEM

It has been found to be best suited for areas receiving <200 mm rainfall, or in degraded rocky-gravelly areas. The highly compatible trees with grasses are: *Acacia senegal, Acacia tortilis, Albizzia lebbek, Tecomella undulata, Colophospermum mopane, Dichrostachys nutans, Hardwickia binnata,, Z. nummularia* and *Z. rotundifolia*. Among the pasture legumes *Clitoria ternatea* and *Lablab purpureus* showed good compatibility with *Lasirus sindicus* and *Cenchrus ciliaris* (Pratap Narain and Tiwari 2005). Under goat grazing, *Z. nummularia* with grass strips in 1:2 ratio led to higher economic returns due to weight gain of the animals and higher wool production (Bhati, 1997a). Under cattle grazing the silvi pasture system with *Z. rotundifolia* and *C. ciliaris* could sustain 554 animal days/ha (Tharparker breed) with 60% pasture utilization (Bhati 1997b).

SOIL AND WATER CONSERVATION

Arid region of Rajasthan receives approximately 50123 MCM water through rainfall. Out of this, 3429 MCM water goes out as basin flow during good years and 501 MCM goes for soil storage (Venkateswarlu et al 1990). Tapping just 10% of this rainfall could cause a perceptible change in drought coping capacity of this region. About 19 m ha area in arid zone is affected by problems of soil erosion by water, wind and various other types of soil degradation such as salinity, waterlogging in canal areas, ravines and mine spoils. Stability in farming can be achieved through improved soil management and conservation technologies like contour bunding, conservation tillage, wind strip cropping, water harvesting and recycling, use of manures and fertilizers, mulching, shelterbelts etc. These technologies proved effective in providing more moisture for crop growth and reducing runoff and soil loss.

INTEGRATED FARMING SYSTEM MODEL FOR ARID ZONE

Success of Integrated farming system lies in proper enterprise mix and optimum utilization of resources. On the above considerations a 7 ha Integrated farming system model has been developed at CAZRI based on twin strategy of system as well as crop diversification (rainfall zone 300-400 mm). This includes arable cropping (15%), agroforestry (35%), agri-horticulture (20%), agripasture (10%) and silvipasture (20%). Crop diversification of pearl millet (40%) with pulses (30%), oilseed (10%) and clusterbean (20%) was found more profitable and sustainable. The perennial components are *Prosopis cineraria*, *Hardwickia binnata* and *Acacia tortilis* in agroforestry, *Ziziphus mauritiana* in agri-horticulture, *C. mopane* and *Z. rotundifolia* in silvipasture. Ten years of study of this model indicates that amongst farming system components B:C ratio was highest in the silvipasture (3.11) and was the lowest in arable farming (1.62) (Fig. 2). The system as per family and herd size was surplus in pulses, fruits, fuel-wood and milk whereas; it was marginally deficient in cereals, fodder and concentrate but substantially deficient in oilseeds. Hence the diversification of farming system components should be in tandem with growing family needs and natural resource sustainability. Over the decade (2001-10) the IFS model had showed reasonable Internal rate of return (IRR, 21.6), net present worth (NPW $ 10560) and B:C ratio (2.05). Besides, a huge amount of carbon was also sequestered in tree components and into the soil (Unpublished data). This system has shown resilience in low rainfall years (2004, 2009), long dry spells of midterm drought (year 2008, 2011), delayed onset of monsoon (2006, 2012) besides recurring terminal droughts. In all such situations, the perennial component like, fruit, fuel wood and fodder trees as well as perennial grasses provided the much needed cash as well as fodder for sustenance of livestock. Based on the experiences of this model and location specific validation through on farm trials ,diversified farming systems are suggested (Table 3)

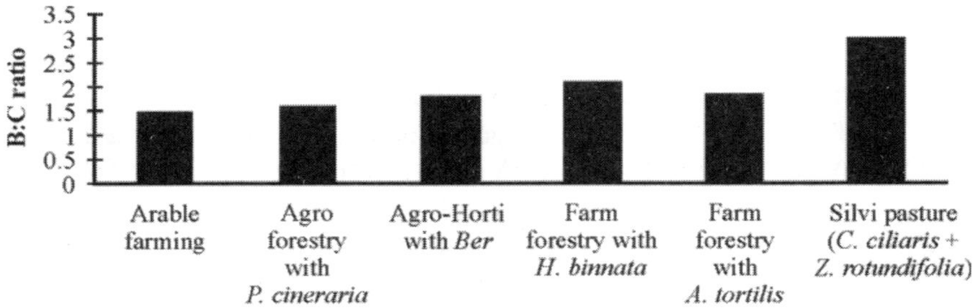

Fig. 2. Benefit cost ratio of various components of IFS model (Average of 10 years)

Table 3. Diversified farming systems for various rainfall zones (% area) and their suitability parameters

	Rainfall situation (mm)			
	<250	**250-350**	**350-450**	**450-600**
Land use (% area)				
Arable cropping	25	30	20	20
Agro- forestry	40	40	40	30
Agri –Horticulture	-	15	30	40
Silvi pasture	35	15	10	10
Livestock sustainability (ACU/year)	1.5-2.0	1.5-2.5	2.5 -3.0	3.0-4.0
B:C ratio	1.5 -2	1.5-2.5	2.5 -3	2.5 -3.5
Net returns (US$)	140-200	200 -400	400- 700	700-1000
Gestation period (Max.- Min)	8-6	6-4	4-3	4-3
Employment generation (man days/year)	200-300	300-500	400-600	500-700

CONCLUSION

Drought is a common feature in arid regions and crop production alone cannot impart stability. The positive aspect of arid zone farming is that by default it is tree- crop farming (agroforestry) at different intensity. Trees moderate microclimate and do not offer competition or sometimes have synergistic effect. During drought or otherwise also, trees provide top feed as fodder, food, fuelwood and some dire needed cash. However, these coping strategies are subsistence in nature. Integrated farming system is an answer to the stability at higher production levels. It amalgamates traditional wisdom with recent scientific learning. Accordingly, integrated farming system models have been developed. These are basically Crop-tree-grass-livestock continuum. The perennial components provide much needed cushion against drought and optimally utilize the scarce resources of arid zone.

REFERENCES

Bhatanagar P, Kaul MK and Singh J 2007 Effect of intercropping in kinnow based production system. *Indian Journal of Arid Horticulture* **2(1)** 15-17

Bhati TK 1997 b Management of dryland crops in Indian arid ecosystem. *In Desertification control in the arid ecosystem of India for Sustainable Development* (eds) Surendra Singh and Amal Kar, Agro Botanical Publishers (India), Bikaner, 298-307 pages

Bhati TK 1997a Integrated farming systems for sustainable agriculture on drylands. *In Sustainable Dryland Agriculture*, Central Arid Zone Research Institute, Jodhpur, 102-105 pages

Bhati TK and Faroda AS 2000 Integrated farming systems for sustained productivity in hot arid ecosystems of India. *In Proceedings of First International Agronomy Congress on Agronomy, Environment and Food Security for 21st Century* (eds) Punjab Singh, Rajendra Prasad and IPS Ahalawat. Indian Society of Agronomy, New Delhi, 342-350 pages

Bhati TK and Joshi NL 2007 Farming systems for the sustainable agriculture in Indian Arid zone. *In Dryland Ecosystem: Indian Perspective* (eds) KPR Vittal, RL Srivastava, NL Joshi, A Kar, VP Tewari and S Kathju, Central Arid Zone Research Institute and Arid Forest Research Institute, Jodhpur, 35-52 pages

Bhati TK, Tewari JC and Rathore SS 2008 Productivity and dynamics of integrated farming systems in western Rajasthan. In *Diversification of Arid Farming System* (eds) Pratap Narain, MP Singh, A Kar, S Kathju and Praveen-Kumar, Arid Zone Research Association of India and Scientific Publisher (India) Jodhpur, India, 23-30 pages

Burman U, Praveen Kumar, Harsh LN 2002 Single tree influence on organic forms and transformation of nitrogen in arid soils. *Journal of Indian Society of Soil Science* **50(2)** 151-158

Dauley HS 1994 Intercropping systems for the Indian Arid zone. *In Sustainable Development of the Indian Arid Zone* (Eds. R.P. Singh and S. Singh,), Scientific Publisher, Jodhpur, 251-259 pages

Faroda AS 1998 Arid Zone Research: An overview. *In Fifty Years of Arid Zone Research in India* (eds) AS Faroda and M Singh, CAZRI, Jodhpur, 1-16 pages

Gupta JP, Rao GGSN, Ramakrishna YS, Rao Ramana 1984 Role of shelterbelts in arid zone. *Indian Farming* **34(7)** 29-30

Gupta JP 2000 Agroforestery for sustained production. In: *Technology Approach for Greening Arid Lands*, Central Arid Zone Research Institute, Jodhpur.

Khan MA, Tewari JC, Isaac VC 1995 Hydrology of small forested catchments in the arid region of Rajasthan. *Annals of Arid Zone* **34(4)** 259-262

McKee TB, Doesken NJ and Kliest J 1995 Drought monitoring with multiple time scales. *In Proceedings of Ninth Conference on Applied Climatology*. Boston: American Meteorological Society. 233-236 pages

Monteith JL, Ong CK and Corlett JE 1991. Microclimatic interactions in agroforestry systems. *Forest Ecology & Management* **45** 31– 44

Oldeman RAA 1983 The design of ecologically sound agroforests. *Plant research in agroforestry* , (ed) Huxley PA. International Council for Research in Agroforestry (ICRAF), Nairobi, Kenya, 173-207 pages

Pareek OP and Awasthi OP 2008 Horticulture-based farming systems for arid region. *In Diversification of Arid farming System* (eds) Pratap Narain, MP Singh, A Kar, S Kathju and Praveen-Kumar, Arid Zone Research Association of India and Scientific Publisher (India) Jodhpur, 12-22 pages

Pratap Narain and Tewari JC 2005 Trees on agricultural fields: a unique basis of life support in Thar Desert of India. In Proceedings of International Conference on Multipurpose Trees in the Tropics: Assessment, Growth and Management organized at Arid Forest Research Institute, Jodhpur during 22-25 November, 2004, 516-523 pages

Ramakrishna YS and Shastri ASRAS 1977 Microclimate under *Acacia tortilis* plantation. Annual Progress Report, CAZRI, Jodhpur, 69-70 pages

Rao KPC, Steenhuis TS, Cogle AL, Srinivasan ST, Yule DF and Smith GD 1998 Rainfall infiltration and runoff from an Alfisol in semi-arid tropical India. II. Tilled systems. *Soil and Tillage Research* **48** 61-69

Rao AS, Surendra Poonia and RS Purohit 2012 Pearl millet yield and drought assessment using standardized precipitation index. *Annals of Arid Zone* **51(2)** 81-84

Rathore VS, Singh JP, Soni ML and Beniwal RK 2008 Response of rainfed aloe to FYM application. *In National Seminar on Opportunities and Challenges of Arid Horticulture for Nutrition and Livelihood.* Organized by Indian Society for Arid Horticulture, Central Institute for Arid horticulture, Bikaner. 54 pages

Saha DK, Amal Kar and Roy MM 2012 Indicators of drought vulnerability for assessing coping mechanism in Arid Western Rajasthan. *Annals of Arid Zone* **51(1)** 1-9

Sharma AK and Gupta JP 2001 Agroforestry for sustainable production under increasing biotic and abiotic stresses in arid Zone. *In Abstracts, Impact of Human Activities on Thar Environment*, Arid Zone Research Association of India, Jodhpur 95-96 pages

Singh KS and Lal P 1969 Effect of *Prosopis cineraria* and *Acacia nilotica* on soil fertility and profile characteristics. *Annals of Arid Zone* **8** 33-36

Singh KC 1989 Importance of grasses in management of sandy soils-strip cropping of grasses. *In International Symposium on Managing Sandy Soils.* Central Arid Zone Research Institute, Jodhpur, 561-564 pages

Singh KC 1995 Management of pasture lands in arid regions. *In Sustainable Development of Dryland Agriculture in India* (Ed. R.P. Singh), Scientific Publishers, Jodhpur 415-423 pages

Singh KC and Gupta JP 1997 Ley farming for sustainable production in arid region of Rajasthan. In: *Proceedings of the Symposium on Recent Advances in Management of Arid Ecosystem*, held at Central Arid Zone Research Institute, Jodhpur, March 3-5, 1997, 95 pages

Singh RS and Kumar A 1993 Agri-horticulture systems under semi-arid conditions. *Agroforestry News* **5(1)** 3-5

Singh RS, Gupta JP, Rao AS and Sharma AK (2003). Micro-climatic quantification and drought impacts on productivity of green gram under different cropping systems of arid zones. *In Human Impact on Desert Environment* (eds) P Naraian, S Kathju, A Kar, MP Singh and Praveen-Kumar, Scientific Publishers, Jodhpur, 74-80

Tewari JC, Harsh LN and Venkateswarlu J 1989 Agroforestry research in arid regions: a review. *In Agroforestry Systems in India – Research and Development* (eds) RP Singh, IPS Ahlawat and G Saran Indian Society of Agronomy, New Delhi, 3-17 pages

Tewari JC, Moola Ram, Roy MM, and Jagdish Chander Dagar 2014. Livelihood improvements and climate change adaptations through agroforestry in hot arid environments. *In Agroforestry systems in India: Livelihood Security & Ecosystem Services, Advances in Agroforestry 10* (eds) J.C. Dagar et al. 155-182 pages

Venkateswarlu JV, Sen AK, Dubey JC, Joshi NL, Kar A, Kolarkar AS, Purohit ML, Ramakrishana YS, Rao AS, Sharma KD, Singh YV and Yadav MS 1990 *Water 200 A.D.: The Scenerio for Arid Rajasthan.* Central Arid Zone Research Institute, Jodhpur, 49 page

WMO 2005 Climate and Land Degradation. WMO-No. 989 2005, World Meteorological Organization.

19

Managing Grazing Lands for Drought Mitigation

J.P. Singh[1], Suresh Kumar[2] and V.S. Rathore[3]

INTRODUCTION

In India, over 40 percent land surface (121 million ha) serves as grazing lands, which have an important place among the forage resources for the huge livestock population. Although largely found in degraded state, they still remain the major forage base throughout the country. They occur in all the agro-climatic zones of the country; but especially they occupy large land areas in the hot arid regions and the cold montane track of the Himalayas. In these regions, they are the backbone of the traditional pasture-based livestock rearing activity. An estimate shows excessive grazing pressure–five times of the normal carrying capacity in the arid regions, three times in the semi-arid regions and two times in the temperate regions (Shankar et al 1996). Besides degradational tendencies, the grassland area is shrinking as crop area and development activities on pasturelands go up and turn intensive.

In the arid region grazing-based livestock husbandry continues to be traditionally predominant occupation and mainstay of the economy and, was found more stable and remunerative than crop farming (Jodha and Vyas 1968, Bhati and Mruthanjaya 1983). Therefore, desert dwellers consider animal husbandry as an instrument of drought proofing (Ghosh et al 1988) and best risk aversion enterprise (Mittal and Prasad 1990, Patel 1991), particularly under rainfed conditions. However, the number of livestock has been increasing and availability of fodder from natural grazing lands has been decreasing. It leads to

---

1. Principal Scientist and Head,
ICAR-Central Arid Zone Research Institute, Regional Research Station, Jaisalmer
2. Principal Scientist, ICAR-Central Arid Zone Research Institute, Jodhpur
3. Principal Scientist & Ex Head, ICAR-Central Arid Zone Research Institute, Regional Research Station, Bikaner
Email: jpsingh@cazri.res.in

the situation of fodder deficit particularly during drought period (Venkatswarulu et al 1992) in the region. Narain (2010) reported that this fodder deficit in western Rajasthan is estimated to be as high as 60% of the demand and might range from 55% in western districts (Bikaner, Jaisalmer and Barmer) to 69% in the central districts (Jodhpur, Nagaur and Churu), and 72% in the eastern districts (Pali and Sikar). Thus during the drought which occurs on an average once in 2.5 year, the scarcity of fodder assumes not merely ecological, socio economic but also geo-political importance. Augmenting the fodder availability from these grazing lands therefore, becomes a priority during normal as well as in periods of drought. Realizing this a lot of research attention has been given to improve the production potential of grazing lands so as to utilize these lands in a sustainable manner. In this paper, an attempt has been made to collate and synthesis this information for improving grazing lands to augment and sustain the fodder availability for drought mitigation in the hot arid region.

GRASSLANDS IN INDIA

Grasslands in the monsoonal tropics are stages intermediate to the final one i.e. forests. Ecologically they represent bio-edaphic climax and not climatic climax. If bio-edaphic pressures are withdrawn they will turn into forests of many types according to the climatic variability. Grass covers in India have been categorized into five types (Dabadghao and Shankarnarayan 1973):

i. *Sehima - Dichanthium;*
ii. *Dichanthium – Cenchrus - Lasiurus;*
iii. *Phragmites – Saccharum - Imperata;*
iv. *Themeda - Arundinella* and
v. Temperate-Alpine

Grasslands above the tree line in the high Himalayas (steppes, alpine meadows) are stabilized by climatic conditions and those in the temperate and tropical zones are of potentially unstable developmental stages of vegetation that developed over times on deforested or abandoned cultivated lands. In the tropical and subtropical parts of India, grasslands are broadly aligned according to climatic and physiographic patterns. These grass covers show close affinity to landforms as habitats and to soil types as niches and have different developmental sequences known as ecological succession. Overgrazing or burning sets off retrogression and such several developmental or seral stages of grasslands are found in India. Range condition surveys conducted in several parts of India reveal that the state of health, in terms of vegetational and habitat parameters, ranges from poor to fair category and grasslands of good or excellent categories are rare. It is because of the tremendous grazing stress on these grasslands. The net result of overgrazing is the loss of desirable species of grasses and habitat degradation by way of erosion.

The hot arid region of India has *Dichanthium – Cenchrus – Lasiurus* grass cover. This grass cover also extends in northern parts of Gujarat, Rajasthan, Aravalli ranges in the south western Uttar Pradesh, Delhi and Punjab. Within this grass cover, there are numerous subtypes attuned to habitat variations. For example, in Rajasthan desert seven potential grassland types on different habitats described by Gupta and Saxena (1972) : (i) *Sehima nervosum* type on the hills and piedmont regions, (ii) *Dichanthium annulatum* type on older alluvial flats with sandy clay loam to clay soils, (iii) *Cenchrus* type on well drained alluvial soils, (iv) *Lasiurus sindicus* type on loose sandy soils, (v) *Desmostachya bipinnata* type on young alluvium, (vi) *Sporobolus-Dichanthium annulatum* type on low lying heavy soils and (vii) *Panicum turgidum* type on sand dunes.

Fig. 1. Oran consisting of wild jujube (*Ziziphus* spp.) woody component, devoid of under storey herbaceous vegetation (village Sodakor, Jaisalmer, Rajasthan)

WAYS TO IMPROVE GRAZING LANDS (*GOCHARS)* AND SACRED WOODLANDS (*ORANS*)

In western Rajasthan, natural grazing lands locally known as *Gochars* and the patches of sacred woodlands called as *Orans* have been the important forage resource base for livestock. The fodder yield from these lands has continuously dwindled due to overgrazing, lack of their management, conversion of these lands to other uses, recurring drought etc. Further, selective grazing by the livestock has resulted in declining the number and density of palatable fodder species. The understorey perennial herbaceous component particularly key grass species of the *Orans* have almost disappeared (Fig. 1). These *Orans* in the arid region are important browse resources for livestock in both normal and the drought period. A number of corrective steps are needed to improve these lands both botanically and habitat-wise. These steps include:

- Auto-recovery of vegetation by way of protection for a specific period through some type of fencing /hedges,

- removal of undesirable bushes,
- adoption of soil and water conservation measures,
- reseeding with desirable species of perennial grasses/legumes,
- planting forage trees/shrubs,
- use of controlled fires,
- cultural operations and fertilizer application.

The grass species of hot arid region like Sewan (*Lasiurus sindicus* Henr.), Dhaman (*Cenchrus ciliaris* L.), Moda Dhaman (*C. setigurus* Vahl), Murath (*Panicum turgidum* Forsk.), Buraro (*Cymbopogon jwarancusa* (Jones) Schult.) are highly tolerant to xeric conditions and well adapted to low rainfall, high temperature and radiation regimes. These are also able to survive the extreme oscillatuions in these climatic parameters including drought. These grasses should be used in reseeding the grazing lands as per habitat.

ROLE OF BROWSE SHRUBS/TREES IN GRAZING LANDS

The leaf fodder or browse from woody perennials (trees and shrubs) is an asset to the livestock, particularly when the grassy vegetation dries up after the monsoon in the arid region. In addition browse also supplements protein to the livestock in deficient grazing lands. In extreme climatic situations of drought and famine browse constitutes the only source of green forage to animals. These woody perennials in grazing lands are important forage resource for small ruminants, particularly goats. In arid region, small ruminants show greater importance in grazing based livestock production. They are vital for resource-poor peasants of arid region to meet their socio-economic needs, thus affording security and survival (Davendra 1998).

Fig. 2. *Haloxylon salicornicum*: an important shrub of arid rangeland

Fig. 3. *Grewia tenax*: showing well developed plant and browsing pressure

In hot arid region, shrubs constitute major source of browse and add a variety to diets of livestock and wildlife because of their palatability, higher level of nutrients and availability in lean period. However, certain shrub species are browed regularly, whereas others at a particular growth stage or under certain seasonal and environmental conditions (Singhand Rathore 2016). Most of the browse shrubs viz., Bordi (*Ziziphus nummularia* (Burm.f.) Wt. & Arn.), Phog (*Calligonum polygonoides* L.), Bawli (*Acacia jacquemontii* Benth.), Gangeran (*Grewia tenax* (Forsk.) Fiori), Lana (*Haloxylon salicornicum* (Moq.) Bunge), Khara lana (*Haloxylon recurvum* (Moq.) Bunge ex. Boiss.) and Gundi (*Cordia gharaf* (Forsk.) Ehrneb. & Aschers. are rich in protein and mineral contents. *Z. nummularia*, that occurs extensively in the scrub grazing lands serves as a leguminous forage by virtue of its high protein content and has a vital role in increasing the carrying capacity of poor scrub grazing lands (Kaul and Ganguli 1963). *H. salicornicum* is an important feed resource for camel and small ruminants (Singh et al 2015). Fruiting tops of *H. salicornicum* contain higher protein (14-19%) and minerals (21-24%) as compared to twigs (Mondal et al 2006) and are well-known camel fodder in the Thar Desert (Fig. 2). Farmers harvest its fruiting top in November-December and preserve for further use as fodder. It could replace the feed concentrate up to 25 and 50% in cattle and goat, respectively (Singh et al 2009, Mathur et al 2012). The *H. salicornicum* together with grass species *Lasiurus sindicus* forms a very productive grazing land in the hot arid region (Shankar and Kumar 1984). *H. recurvum* another chenopod, succulent halophytic shrub is a potential alternate forage particularly during the summer period in rainfed arid lands (Rathore et al 2012). Its green foliage is rich in crude protein (13.2%), total carbohydrates (62.9%), crude fiber (20.7%), neutral detergent fiber (58.2%), and acid detergent fiber (25.2%). Its utilization by mixing with other fodder revealed that it could be fed to goats by mixing with groundnut haulm in 1:3 ratios (Mondal et al 2005). *G. tenax* is heavily browsed by goats in gravelly-rocky rangelands (Fig. 3). In Jaisalmer district of Rajasthan, natural patches of *C. gharaf* occurs in shrubby form, and are very important for livestock particularly small ruminants and for wild animals (Fig. 4).

Fig. 4. Shrub form of *Cordia gharaf* in arid rangeland (village Mehroli, Jaisalmer)

Leaf fodder and other products from fodder trees are important forage resources for livestock during normal and drought period in arid region. Lopping of Khejri (*Prosopis cineraria* (L.) Druce) is practiced in most parts of western Rajasthan and the threshed dried leaves locally known as "*Loong*" are used as fodder. The *Loong* is nutritive and given with concentrate to milch animals for raising milk production. Jal (*Salvadora oleoides* Decne.) is also one of the important tree species of gravelly rocky areas in the hot arid region particularly in Jaisalmer district of Rajasthan (Fig. 9). Besides its browse value, its fruits locally called as "*Pilu*" are also fed to animals especially milch cows and supposed to increase the milk production.

ADOPTION OF GRAZING MANAGEMENT FOR SUSTAINABILITY OF GRAZING LAND

Animal grazing has significant impact on grazing lands as it affects both production and form of plants. Long periods of continuous grazing often lead to deterioration in composition and yield of good forage grasses and increases in the unpalatable species (Fig. 5). Overgrazing of grazing lands results in slow but an inexorable deleterious change in botanical composition, through, it varies with the type of grass cover. Again in the hot arid region, plant growth is limited to a short period of moisture availability due to eco-climatic conditions; resulting in often large inter-annual variation in forage production from the same grazing lands. This situation of grazing lands is further accentuated during drought. Therefore, a well-managed grazing system needs to be followed to provide optimum nutrients on sustained basis to the livestock with lesser negative environmental impacts. Further, efficient use of grazing lands can only be made, if the duration and intensity of animal grazing is regulated properly. The grazing pressure in arid areas is very high i.e. 1 to 4 ACU/ha against the normal 0.2 to 0.5 ACU/ha. This can be determined by grazing according to the carrying capacity of the grazing land.

Fig. 5. Effects of grazing management in arid grazing lands : (a) Protected (fenced) *Lasiurus sindicus* dominated farmer's own grazing land with regulating grazing (b) free range grazing land consisting of decreased cover of *L. sindicus* and infested with bushes at Chandan, Jaisalmer, Rajasthan

Results of various grazing experiments indicate that deferred –rotational grazing system helps to maintain the range health. In the deferred system, the grazing area is divided into compartments, and at least one of these is rested until seed-setting. In the rotational grazing no compartment is rested and so all of them are grazed in rotation of specific duration. The deferred-rotational system is a mix of the above two types and is considered as the best system of grazing to improve the quantity and quality of pasture from grazing lands. Selective grazing can also be avoided by practicing this system of grazing. Further, types of animals to be grazed depend on the kind of grazing lands for sustainable utilization of available biomass. For example, mixed herd (like cattle, sheep and goats) grazing is recommended in shrub-dominated grazing lands, because forage choice of cattle, sheep and goat are distinct and *inter alia* complimentary to maintenance of vegetation composition (Singh et al 2001).

UTILIZATION OF WASTELAND FOR FODDER RESOURCE

Sand dunes: About 68% of the area of western Rajasthan is occupied by sand dunes and sandy plains, which need major focus in forage development programme. Sustainable utilization of sand dunes for growing of suitable species (Grass: Sewan (*Lasiurus sindicus*) and Murath (*Panicum turgidum*); shrubs : Phog (*Calligonum polygonoides*), Bawli (*Acacia jacquemontii*) and also Lana (*Haloxylon salicornicum*) will increase the fodder production along with maintaining the ecological balance. The *C. polygonoides-Panicum turgidum* are the most common plant community of psammophytic scrub desert which occurs on all three types of dunes of the old system. *C. polygonoides* which is an important climax species of sand dunes is drought hardy, deep-rooted and has high regeneration capacity (Fig. 6). It is commonly browsed by goat, camel and sheep and eaten by all animals in scarcity and drought. Further, this species serves as best check of soil erosion in sandy areas. Further, the establishment of shrub species like *H. salicornicum* and *C. polygonoides* alter physical soil traits and enhance soil nutrients; and also facilitate the development of surrounding herbaceous plant communities (Rathore et al 2015). *P. turgidum* is a very drought hardy species, and possesses an excellent sand binding capacity (Fig.7). Likewise, *A. jacquemontii* is another climax leguminous shrub of sand dunes, which is preferably browsed by camel and goats (Fig. 8). Hence, there is a need to conserve and promote such type of key species. In fact, farmers should be motivated to utilize dunes in their agricultural fields for site-specific silvi-pasture systems.

Fig. 6. *Calligonum polygonoides* : a major shrub of arid sand dune grazing lands (a) pure stand and (b) in association with *Lasiurus sindicus* (Jaisalmer, Rajasthan)

Fig. 7. *Panicum turgidum* : an important grass (a) in sandy undulating plains (b) dune area of arid grazing lands

Fig. 8. *Acacia jacquemontii* : an important shrub species for sand dune fixation and as source of browse

Fig. 9. *Salvadora oleoides*: a multipurpose species of arid rangelands and provides browse, shelter and edible fruits

Shallowly gravelly, rocky and stony wastelands: The adapted browse woody perennials like Gangeran (*Grewia tenax*), Gondi (*Cordia gharaf*), Jal (*Salvadora oleoides*) etc., have important role in providing the browse material to small ruminants in gravelly, rocky areas (Fig. 9). *G. tenax* is one of the drought hardy species that bears immense browsing pressure in hot arid region. Guggal (*Commiphora wightii* (Arnott.) Bhandari), which is one of the important threatened medicinal species occurring naturally on open dry hills and rocky habitats in Jaisalmer district. However, it is browsed by the livestock and wild animals and regenerates very fast after receiving the rainfall. It also protects many wild species of fodder importance in the rangelands. Gupta et al (1998) reported that contour furrow trenching with planting of Kair (*Capparis deciduas* (Forsk.) Edgew.) and Bordi (*Ziziphus rotundifolia* Lamk.) for management of shallow gravelly and stony wastelands which cover ~1 m ha in Jaisalmer and Barmer district. Similarly on rocky and gravely wastelands, the planting of shrubs like *Z. rotundifolia* performed very well. *C. decidua* an important source of browse to small ruminants during lean period and drought situations is browsed eagerly by camel and goats (Fig. 10) particularly during post winter and provide shelter to livestock and wild animals.

Fig. 10. *Capparis decidua*: a shrub species of arid gravelly rangeland and provides browse to camel and goats.

Fig. 11. Sweet saline depression (Meetha ran) area in hot arid region (Jaisalmer, Rajasthan) comprised mainly true halophyte species *Aeluropus lagopoides* and *Cressa cretica.*

***Salt affected lands*:** Salt affected wasteland areas are barren or having adapted halophytic vegetation in pockets. Suitable halophytic grasses and shrubs offer good scope of providing nutritious forage to animals particularly goats and camel. For example, in Tal Chhaper (Churu district), Khara ghas (species of *Chloris* and *Sporobolus*) grow well naturally and produced considerable green forage along with chenopod shrub Luni (*Suaeda fruticosa* (L.) Forsk.). In Meetha ran in Jaisalmer having the good cover of true halophyte species viz., *Aeluropus lagopoides* (L.) Trin. ex Thw. and *Cressa cretica* L. are grazed by livestock (Fig. 11). These problem lands, if utilized properly hold promise to increase the forage production by planting the salt tolerant species of grasses, shrubs and trees. The halophytic shrubs like *Haloxylon recurvum, Salsola baryosma* (Roem. ex Schult.) Dandy, *Suaeda fruticosa* and the grasses like *Sporobolus ioclades* (Trin.) Nees, *Chloris* spp, have the promise and need desired attention. Most of the halophyte species form the important forage resource base for camel and small ruminants in hot arid region.

PROMOTION OF TRADITIONAL GRAZING AND FODDER CONSERVATION PRACTICES

Disruption of traditional grazing system is also one of the causes of grazing lands degradation. There is a need to promote the traditional grazing practices of the past, for sustainable grazing resource management. Storage of surplus fodder has been prevalent since ancient times in order to feed the livestock during the normal and lean period. Annual grasses like Bhurat (*Cenchrus biflorus* Roxb.) that commonly occurs throughout the arid region in grazing lands, field bunds and agricultural fields are harvested and stacked. Some of the perennial grasses like Sewan (*Lasiurus sinducus*), Dhaman (*Cenchrus ciliaris*), Moda Dhaman (*C. setigerus*), Ganthia (*Dactyloctenium scindicum* Boiss.), Khari ghas (*Sporobolus* spp.) etc. are harvested and show high stacking period. Other wild legumes like Bekaria (*Indigofera cordifolia* (Heyne ex Roth) is collected and stacked for stall feeding during the lean period and fed in mixture with other feeds (Fig. 12). The wild species viz., Bakda (*Tribulus pentandrus* Forsk. Syn. *T. alatus* Delile) is also stacked and fed to camels, sheep and milch cattle during the lean period. Leaf fodder of perennials like *Pala* from Bordi (*Ziziphus nummularia*), *Lasu* from Phog (*Calligonum polygonoides*), *Fulli* (fruiting tops) from Lana (*Haloxylon salicornicum*), *Loong* from Khejri (*Prosopis cineraria*) etc are traditionally stored. Such fodder conservation practices afford resilience thus help in mitigating drought.

ESTABLISHMENT OF FODDER BANKS

Establishment of fodder banks have great relevance as they meet the demand during lean period as well as during drought. Relatively inaccessible areas having drinking water problem such as those in Jaisalmer district, have a surplus of fodder (Shankar and Kumar, 1987) which can be harvested and conserved/preserved in fodder banks for drought period. There is a need to identify such pockets in the region for sustainable utilization of available biomass. Moreover, in Jaisalmer district, a considerable area of rangelands is under the arid shrub Lana (*Haloxylon salicornicum*). Its fruiting tops locally called Fuli are traditionally harvested for livestock feeding. If proper harvesting at fruiting stage is carried out from its natural stands, it may be a good source of nutrients to livestock. Dune fields in the Bikaner and Jaisalmer districts have some very good patches of Phog (*C. polygonoides*), that provide leaf fodder, *Lasu* and the fruits called as *Ghintal,* the rich source of nutrients to animals particularly camels. Collection of leaf, young twigs and fruits from existing areas, and preserving them in fodder bank provide a strategic edge in mitigating drought. Grasses from the protected areas of the forests and also from other protected areas where grazing is restricted can also be utilized for these fodder banks. Thus fodder banks can be established at the focal points utilizing the fodder resources from the fodder sufficient sites.

Fig. 12. Stacked dried fodder *Indigofera cordifolia :* an important dried fodder for camel in gravelly sandy areas of arid region (Pokran tehsil, Jasisalmer, Rajasthan)

CAPACITY BUILDING OF STAKEHOLDERS

Though technologies in the grazing land management and in related aspects are available; their dissemination in a participatory mode in the field is far below expectation. Generally, less attention is given in pasture development/grassland management aspects in farmer's trainings and other extension activities. Focus should be given to improvement of degraded *Gochars*/grasslands, improvement of *Orans* (scared woodlands), grazing management, promotion of silvi-pastures, rangeland management, rehabilitation of various types of wastelands etc. Recently, Singh et al (2016), stressed that "It is important to involve and aware the desert inhabitants regarding the benefits of rangeland development and management techniques in order to achieve sustainable livestock production and upkeep of rangeland health and conservation of biodiversity". Strengthening and capacity building of various stakeholders in the field of grazing land management and related aspects will surely help to build a strong fodder resource base so as to mitigate the impact of drought.

AWARENESS CAMPAIGN AND ACTIVE PEOPLE PARTICIPATION

It is of utmost importance to create awareness amongst the herders/farmers regarding the value of grazing lands and their sustainable utilization particularly in regard to range use. It will certainly ensure the sustainability of *Gochars/ Orans* and precious rangelands of the region. Many of the adapted browse shrubs from the grazing lands are over exploited and people even dig out their roots for fuel and other purposes. In view of harsh climatic conditions, regeneration of overgrazed grazing lands is really a problem; however, physical removal is of greater danger. Therefore, awareness should be created not to dig out the roots of precious browse species which support the livestock during lean periods and drought and hence, vital for arid environment. There is a need of active people/community participation and to form a Village Grazing Land Management Committee (VGLMC) for overall management of village grazing lands.

CONCLUSION

Grazing lands are vital for sustainability of livestock based economy of arid region. State of these grazing lands varies due to topographic and climatic situations, so location specific impact assessment is essential. The productivity of these existing grazing lands can be increased by adopting available technologies and replacing low yielding annual grasses with high yielding nutritive adapted perennial grasses. In degraded grazing lands renovation or rehabilitation programme, highly drought tolerant perennial grasses, shrubs, and some succulents should be used which characterize the desert ecosystem. In addition to this emphasis should also be given in adoption of improved soil and water conservation measures in grazing lands, which is very much required in the arid region. Further, increasing the diversity in the vegetation composition and in the phenology of component species in grazing lands will ensure richness in forage quality and availability of biomass over a longer but staggered point of time. However, utilization of existing grazing lands should be based on carrying capacity assessment.

Creation of awareness amongst the herders/farmers in regard to value of grazing lands is very much desired. *Orans* (sacred woodlands), are important browse resource base in western Rajasthan during the drought situation. They should not be projected as the browse forage resource base but also as a heritage site of *in situ* conservation of hot arid biodiversity. Emphasis should also be given on strengthening the extension services in the areas of grazing land management and related aspects. There is also a need to promote the traditional fodder conservation practices in the region. Ensuring active community participation for sustainable utilization of grazing lands and adoption of suitable grazing systems is very important and need of the hour. Moreover convergence of different government and non-government departments is vital in managing the grazing land resource base for sustainable utilization and to mitigate the impact of drought.

REFERENCES

Bhati GN and Mruthanjaya 1983 Economics of high yielding perennial pasture grasses in arid land of western Rajasthan. *Forage Research* **7** 173-180

Dabadghao PM and Shankarnarayan KA 1973 *The Grass Cover of India*. ICAR, New Delhi.

Davendra C 1998 Improvement of small ruminant production system in rainfed agroecolgical zones of Asia. *Annals of Arid Zone* **37** 215-232.

Ghosh PK, Ojha DC and Bohra HC 1988 Role of livestock in the socio-cultural milieu of the Rajasthan desert. *In Integrated Socio-ecological Development of Desert.* CAZRI, Jodhpur. 1-13pages

Gupta JP, Aggarwal RK, Joshi DC, Sharma KD and Singh KC 1998 Management of wastelands for sustainable production. *In Fifty Years of Arid Zone Research in India* (eds) AS Faroda and M Singh CAZRI, Jodhpur (India)153-166 pages

Gupta RK and Saxena SK 1972 Potential grassland types and their ecological succession in Rajasthan desert. *Annals of Arid Zone* **11** 198-218

Jodha NS and Vyas VS 1968 *Growth and Stability in Arid agriculture*. Agro-economic Research Centre, Vallabh Vidyanagar, Gujarat.

Kaul RN and Ganguli BN 1963 Fodder potential of *Zizyphus* in the shrub grazing lands of arid zones. *Indian Forester* **89** 623-630

Mathur BK, Singh JP, Rathore VS, Singh NP and Beniwal RK 2011 Utilization of hot desert shrub: Lana (*Haloxylon salicornicum*) seeds as feed resource in arid zone. *Indian Journal of Small Ruminants* **17** 231-234.

Mittal JP and Prasad S 1990 Place of dairy farming in arid agriculture. *Indian Dairyman* **42** 17-18

Mondal BC, Singh, JP and Beniwal RK 2006 Chemical composition of desert shrub Lana (*Haloxylon salicornicum*) used as animal feed in western Rajasthan. *Annals of Arid Zone* **45** 91-92

Mondal BC, Singh JP, Beniwal RK and Rathore VS 2005. Palatability of Khara Lana (*Haloxylon recurvum*) in goats. *Indian Journal of Small Ruminant* **11** 219-220.

Narain P 2010 Feed and management in arid region. pp. 184-197. In *Feeding and Management of Livestock during drought and Scarcity* (eds) NV Patil, BK Mathur, AK Patel, M Patidar and AC Mathur, Scientific Publishers, Jodhpur

Patel RK 1991 Role of livestock production in rainfed agriculture: a comparative study, *In Proceeding of the International Conference on Technologies for Minimizing Risk in Rainfed Agriculture*, Indian Society of Agricultural Extension Education, New Delhi, pp. 394-411.

Rathore VS, Singh JP and Roy MM 2012 *Haloxylon stocksii* (Boiss.) Benth. et Hook. f., a promising halophyte: distribution, cultivation and utilization. *Genetic Resources and Crop Evolution* **59** 1213–1221.

Rathore VS, Singh JP, Bhardwaj S, Nathawat NS, Mahesh Kumar and Roy, MM 2015 Potential of native shrubs *Haloxylon salicornicum* and *Calligonum polygonopides* for restoration of degraded lands in arid western Rajasthan, India. *Environmental Management* **55** 205-219

Shankar V and Suresh Kumar 1987. Ecological distribution of *H. salicornicum* (Moq.) Bunge as an aid to the reconstruction of the lost courses of the Saraswati River in the Indian Desert. *Tropical Ecology* **25** 227-238

Shankar V and Suresh Kumar 1987 Grazing Resources of Jaisalmer- Ecology and Developmental Planning with special reference to sewan grasslands. CAZRI, ICAR, Publ. No. 1-92.

Shankar V, Gupta, JN and Singh JP 1996 Forage Production technology on rangelands. Pp. 41-51. *In Transfer of Forage Production Technologies* (eds) Bhag Mal and R.J. Hagger, Indo-UK Forage Production Project, IGFRI, Jhansi.

Singh JP and Rathore VS 2016 Shrubs of hot arid region: Diversity, utilization and conservation. *In* Improving *Productivity of Drylands by Sustainable Resource Utilization and Management* (eds) Devi Daval, D Machiwal, S Mangalassery and RS Tripathi. New India Publishing Agency, New Delhi. 235-252 pages.

Singh, JP, Mathur BK and Rathore VS 2009 Fodder potential of Lana (*Haloxylon salicornicum*) in hot arid region. *Range Management & Agroforestry* **30** 34-37.

Singh, JP, Rathore VS and Roy MM 2015. Notes about *Haloxylon salicornicum* (Moq.) Bunge ex Boiss., a promising shrub of arid regions. *Genetic Resources and Crop Evolution* **62** 451-463.

Singh JP, VS Rathore and Bhatt RK 2015 Hot arid rangelands: Forage resource base utilization and conservation. *In Grasslands: A Global Resource Perspective* (eds) P K Ghosh, SK Mahanta, JB Singh and PS Pathak, Range Management Society of India, Jhansi, India 63-82 pages

Singh JP, Vinod Shankar and Pailan GH 2001 Managing bush infested rangelands through mixed herd grazing. In *Impact of Human Activities on Thar Desert Environment. Symposium Abstracts*, Feb. 15-17, 2011. CAZRI, Jodhpur.

Venkateswarlu, J, Anantharam, K, Purohit, ML, Khan, MS, Bohra, HC, Singh, KC and Mathur, BK 1992 *Forage 2000 AD: The Scenario for Arid Rajasthan*. CAZRI, Jodhpur.

20

Livestock Management in Drought

A.K. Misra

INTRODUCTION

Livestock production is an integral part of dryland agricultural system in India. It plays multiple roles in rural systems and economy and has a strong human dimension, as manifested through socio-cultural link and involvement of women (Rangnekar 2006). Besides its well-established role in agriculture, livestock plays a crucial role in food security and as risk aversion mechanism for sustaining family, whenever there are crop failures (Misra et al 2006). Livestock enables poor and landless farmers to earn income using common-property resources (World Bank 1999), and provide a constant flow of income and reduce the vulnerability of agricultural production (Holmann et al 2005).

The nature has endowed dryland areas with some of the best breeds of cattle, sheep, goats and other species of livestock. It is also endowed with nutritious perennial grasses. Failure of rain affects directly the growth of these grasses as well as water resource to the livestock. The immediate impact of drought is on the health of livestock, due to lack of feed, fodder and drinking water. In drought of 1987, about 120 million out of 214 million livestock population in 17 drought-affected states, were affected severely (Negi 2002). It was estimated that the drought of 1987 killed almost one-third of cattle in these areas (Bhasin 2002). The impact of drought on livestock population of Barmer district of Rajasthan is presented in Table 1. Large-scale migration of livestock- mainly cattle and sheep take place from these areas, resulting in hardship to livestock farmers besides attracting a host of social problems. Most of the areas not affected by drought remain unwilling to receive these livestock since they may damage their crop. There are invariably clashes between the local farmers and farmers migrating with their livestock (Singh 2002). Feed scarcity, during drought years, is a common feature in arid and semi-arid regions of India. Livestock-keepers require

Principal Scientist and Head
Central Arid Zone Research Institute, Jodhpur 242003, India
Email:arun.mishra@icar.gov.in

a supply of forage for their animals throughout the year, and utilize the resources available to them in ways that minimize feed scarcity. Farmers obtain forage from a combination of crop residues, private grazing land (not used for crop production), common grazing land and purchased fodder. The response of farmers to the decline in feed and fodder resources during drought has been three-fold: (i) they attempt to maximize the complementarity between common and private property resources; (ii) they accept inferior options (feed resources) because of a lack of alternatives; and (iii) they share feed resources (weeds and crop residues) with labourers in the form of in-kind payment and neighbors. Ruminants, owing to their larger body size, need huge quantity of fibrous feed material. However, roughages are usually not available in adequate quantity during droughts. Concentrates can be purchased or brought from other areas. Therefore, during scarcity, it is important to find out the alternative feed resources as roughage substitutes, which will be helpful in providing minimum roughage requirement of animals.

Table 1. Impact of drought on livestock population in Barmer district of Rajasthan

Species	Population ('000)		
	Pre drought, 1983	**Post drought, 1988**	**Per cent decrease**
Camel	109	94	13.76
Buffalo	41	35	14.63
Goat	1335	743	44.34
Cattle	491	234	52.34
Sheep	1157	482	58.34
Total	3184	1682	47.17

Source: Livestock Census, Govt. of Rajasthan, 1989

INCREASING FEED RESOURCE BASE

The following measures are suggested for increasing the feed resource base during drought periods (Patil et al 1979, IGFRI 1987, Shukla et al 1988).

Prolonging the life of existing crops: In order to get maximum yield from the existing crops, it is necessary to reduce competition between plants for moisture and essential nutrients. The crop population be immediately thinned out and reduce to 40 to 50% of the original population level. The excess of vegetation thus available should be collected, chaffed and then fed to the animals. If there remains some surplus forage material, this should be properly processed and fed to the animals.

Use of failed crops as animal feed: Whenever failure of monsoon occurs, the existing crops fail to grow adequately and such crops can be utilized for feeding the animals during drought conditions.

Increasing grass production in the fallow lands, forest areas and Govt. farms: It has been observed that in the drought year, a large area remains fallow. Under such situation, it is advised that farmers may apply any nitrogenous fertilizer @ 20 kg/ha on the fallow land or community grazing land. This will help in two ways: (a) it will increase the growth of grasses if there are some unscheduled rains, and (b) it will improve nutritive value of grasses. In the forest areas as well as the Govt. farms, it is advised that the available grasses be harvested and after harvesting, urea may be applied in order to increase faster re-growth and more biomass production.

Fodder production on banks of stagnated water ponds, rivers, canals, nalas and empty tank beds: As an emergency measures these resources should be tapped for fodder production, even on temporary basis if not on permanent basis. Grasses like Hybrid Napier, Guinea, Setaria intercropped with fodder shrubs like *Sesbania, Leucaena, Glyricidia,* should be planted in the deep Central and Southern India and *Sarson, Vetches, Senji, Barley* in tank beds in the North India. Sewage water may be used for fodder production nearby cities. Para grass, Hybrid Napier and guinea grass may be grown as perennial fodder sources (Singh 2002).

Collection and use of available feedstuff: All kind of edible green herbage including tree leaves available from grassland and forest areas and seasonal fallow lands be immediately harvested, dried, baled and stored for immediate feeding and future use. Harvesting of these herbage materials at proper stage will result in 50-80% higher production of nutrients and dry matter. Apart from these, large quantities of shrubs and tree leaves, withering crops as well as aquatic plants such as water hyacinth, para grass, singhara, algae and other under-water vegetation could be collected and fed to the animals. In addition, agro-industrial by-products such as fruit pomage, tapioca starch waste, and damaged grains from warehouses, etc. could be fed to the animals in the drought stricken areas. Non-traditional unconventional feed resources such as seeds of some trees as *mahua, sal, cassia tora, tamarind and subabool and pods of Prospis juliflora* and *Acacia nilotica* etc. could be collected and processed for feeding to the animals. If these materials are collected and processed, particularly during drought year, they could serve an important source of animal feed.

PROCESSING OF FORAGES FOR EASY TRANSPORT

The following technologies are recommended for processing of forages for easy transport during drought periods (IGFRI 1987).

Chaffing: Chaffing of fodder should be popularized in drought prone areas. The advantages of feeding chaffed feed are that it avoids wastage and prevents selective consumption. Feeding of chopped roughage reduces the energy wasted while chewing, increases the feed intake and improves digestibility. The net biological value of the feed also improves (Singh and Prasad 2002).

Hay making: Harvesting, sun drying in swath for 3-4 days followed by one turning and stacking around a tripod frame on raised platform is the simplest and cheapest method of hay making (IGFRI 1987). The low moisture content of hay considerably reduces costs and efforts involved in transportation and handling. The surplus green grasses available during the monsoon in the forest go unused and could be put to the best possible use by this process.

Bailing: Bailing reduces volume of loose grasses to 2.5 times. This makes transportation and handling economic and convenient. Manual operated or power operated hay balers can be utilized for this purpose.

Densification: Compound herbage feed mixture viz., straw 50 to 60%, leguminous crops/tree leaves 35 to 40%, and molasses 10 to 15% or a mixture of leguminous crops and tree leaves, can be densified to the extent of 250 to 300 kg/m^3 using wafering machine. Densified fodders are 6 to 7 times less in volume and are easy and economical to transport to longer distances.

IMPROVING QUALITY AND UTILIZATION OF FODDER

The following technologies are recommended for improving the quality and utilization of roughages (Ranjhan and Khera, 1976, Shukla et al 1988, Singh and Schiere 1995, Singh and Parasad 2002).

Enrichment of low-grade roughages with urea treatment: Low grade roughages such as paddy straw, wheat bhusa, sorghum kadabi, maize stover, dry grasses and other edible farm waste contain negligible amount of digestible crude protein and higher amount of non-digestible cell wall constituents. Through urea treatment, nutritional value of poor quality herbage could be improved and made more palatable. Four percent urea, 50 litres water per 100 kg straw/herbage and 15 to 21 days incubation period are optimum for treatment. Urea treatment is quite flexible, as it can be adapted to local conditions. Urea treatment, apart from being a source of nitrogen for microbial synthesis, also provides additional energy due to the weakening/loosening of the lingo-cellulose bonds in the treated straw. Treatment improves dry matter intake by 7 to 10 units and digestibility by 4 to 15 units. For daily feeding required quantity of treated forage be removed from the heap and fed to the animals.

Urea-Molasses mixture: Low-grade roughage/fodders can be fortified with molasses, urea, salt and mineral mixture. Urea-molasses mix contains soluble and fermentable nitrogen from urea, highly fermentable energy from molasses, and essential minerals.

Preparations: The mixture consists of: Water-1.5 kg; Urea-1.5 kg; Molasses-10 kg and Salt- 1.0 kg. Dissolve 1.5 kg urea in 1.5 kg water. Stir till the urea is dissolved. Add 10 kg molasses and mix thoroughly. Add 1 kg salt and 1 kg mineral mixture. The mixture can be fortified with vetablend @ 25 g per 100 kg liquid feed and can be stored in a covered earthen pot for a period of one month.

Feeding of urea-molasses mixture: Take 0.5 kg of the above mixture each day and mix with 2 kg of water. Sprinkle the solution on at least 5 kg of straw and turn thoroughly. Unless already chopped, all straws must be reduced to 2-3 inches. After 20 days, the mixture may be increased to 0.75 kg of mix to 2 kg of water and used as above. The urea/molasses/salt, mineral mixture can also be used with in a concentrate mixture or with chaffed green fodder. The urea mix must be thoroughly and evenly mixed.

Precautions: The urea mixture should not be fed to animals less than 4 months of age. Animals should have started ruminating before they are fed the mixture. The mixture should not be kept for more than one month.

Urea molasses mineral block (UMMB): It is commercially available and can be used to supplement the low quality roughages in place of urea-molasses mix to balance the deficient nutrients in the ration. Similar to urea-molasses mix, block contains soluble and fermentable nitrogen from urea, highly fermentable energy from molasses, and essential minerals. Natural protein sources such as groundnut or cottonseed extract have also been added to provide preformed peptides and amino acids. UMMB has been found to improve the dry matter intake of the basal roughage and the feed digestibility. The nutrients from the block are well utilized by the animals and UMMB supplementation improves reproductive performance of livestock due to enhanced availability and utilization of nutrients, particularly micronutrients.

Strategic supplementation: Supplementation of straws with deficient nutrients such as nitrogen and minerals to improve intake and digestibility of straw is called catalytic supplementation. Use of small amounts of concentrate, licks, or green fodder may achieve this objective through improved rumen function. Thus, supplementation of available forages with green leguminous forages/ concentrates can considerably increase the livestock productivity.

LIVESTOCK FEEDING AND MANAGEMENT PRACTICES

Under drought condition, three important feeding strategies need to be adopted depending on the severity of the drought situation viz.,

- Feeding strategy for maintenance of animals to ensure their survival
- Feeding strategy to maintain minimum level of production and growth
- Feeding strategy to maintain desired production by purchasing supplements from outside

Critical body weight for survival will be up to 20% of body weight loss in case of cattle, while in case of desert animals (sheep and camel) it may be up to 30 to 40%. In case of milch and other producing animals, the only solution lies with supplementation because productivity loss cannot be recovered by any other means. The important management practices, which are to be adopted during drought condition are:

- Animal should not be kept in direct sunlight. They should be tied under a tree or in a shed. They should be allowed to graze only in early morning or late evening.
- Feeding should be done during early morning and in the late evening, as animal's water requirement will decrease.
- There should be no overcrowding of animals in shed and animals should be provided proper ventilation. If the roof of the animal shed is of iron sheet it should also be covered with thatch or asbestos sheet so that animal may be protected with environmental temperature.
- Upper layer of *Katcha* floor should be replaced with new sand in every six month so that load of organism can be reduced. Pakka floor should be properly cleaned.
- There are certain diseases, which are more common during drought period and hence need, more attention so as to prevent its outbreak. The most common diseases are foot and mouth disease, hemorrhagic septicemia, black quarter, anthrax, enterotoxaemia, ring worm, ascariasis, tick infestation and mangee, etc. To control and prevent these diseases, regular vaccination and deworming schedule should be followed.
- Awareness among the local community must be created, so that they take proper care of their animals and prevent them from eating poisonous toxic material, which may further increase the stress factor.
- During drought, there is heavy mortality in animals due to heat stroke, dehydration, infectious and non-infectious diseases. As infected animals may spread the diseases to other animals, hence proper disposal of the carcass is essential. Burying or burning method of carcass disposal may be adopted.

TACKLING THE DROUGHT: AN ACTION PLAN

A long-term strategy is required to tackle the drought and drought proofing of drought prone areas of the country. This requires diversification of agriculture on the one hand, while feed resources planning on the other. The feed resource planning should include forage and grass production and fodder banking. Fodder should be transported from the surplus areas and stored in the drought affected areas for use during the bad years. At present, most of the by-products of crops are allowed to go waste in agriculturally developed irrigated areas. The irrigated areas do not have initiative to save and store these resources. The strategy should be to transport these resources to the drought prone areas.

The management of drought would primarily be the responsibility of concerned state governments. Every state should formulate its strategy to deal with the draught. However, looking to the extent and severity of the draught in the country, an action plan should be prepared by the government of India to ensure coordination between different ministries and states. The action plan should have both short and long-term measures.

Short term and immediate measures

- Highest priority should be accorded to augmentation of feed resources especially fodders. The provision of fodder and concentrate feed to meet extra feed requirements and mineral-vitamin blocks to balance livestock rations should be made. The government should immediately allocate maximum possible quantity of molasses for livestock feeding. Similarly, the Food Corporation of India is carrying huge quantity of grains unfit for human consumption. These should be released to feed compounding plants especially those owned by cooperatives of livestock farmers for production and supply to drought-affected areas.
- The fodder banks need to be established at a central place, well connected by road and having good source of water supply for immediate use. The coarse roughages and crop residues may be transported to the bank where these can be treated/fortified with deficient nutrients to form a good livestock ration. The material can be stored for a considerable period. The transportation can be made easy by briquetting/ wafering the material to reduce the volume by 6 to 7 times.
- In the drought-affected areas, livestock become prone to various infectious and contagious diseases. Provision of vaccines to cover possible outbreaks of stress-induced diseases with a training package for vaccinators should be made in the drought-prone areas. Animal health programme should also include deworming of growing stock.
- Large-scale relief camps should be organized for livestock near a source of water or where water can be easily provided. In these camps, livestock feeding programme based on urea molasses feed should be taken up.
- Livestock farmers in the affected area would require an efficient marketing network. During drought many milk producers cooperative become inactive and close down their operation. Their milk transportation cost becomes high, making milk collection unviable. They should be provided with subsidies to meet higher cost of transportation and organize distributing of inputs to the farmers. Similarly, special efforts should be made to organize marketing of surplus animals especially sheep and goats to avoid their distress sale. This would require establishment of holding centres in the affected areas and their transportation to major consumption centres.
- Livestock farmers should be given credit for purchase of feed and fodders and other support services on liberal terms.

Long-term measures

It is suggested that various departments should come out with long-term measures to combat drought situations in future to save the livestock. There is a necessity for establishment of animal feed security system at the national level on the pattern of buffer stock of food grains. The long-term measures should have following elements.

- To meet the diverse vagaries of climate, it would be better if such situations do not surprise us and we are prepared in advance to meet the challenge. This may perhaps be achieved through the establishment of feed and fodder banks. These banks can be established in areas of large production of straws, stover, bagasse, so that there is no difficulty in transportation of the raw material for treatment/fortification/ processing. Few pockets may be identified in each state where enough storing facilities can be easily created. In due course of time, they may serve as buffer stock to be transported to the required place to provide immediate relief during the scarcity period which otherwise, is a difficult proposition at present.
- A national network should be created to supply fodder seed of drought tolerant varieties. The animal husbandry departments in the states should develop pilot plants for treatment of roughages and crop residues for animal feeding. Each state with high probability of drought may create infrastructure down to the block level for feed resource development in terms of both, augmenting forage production and processing, and storage of other non-conventional feed materials.
- The degraded forest lands in states may suitably be developed as potential forage growing areas to supplement feed resources. Development of three-tier-agroforestry system of forage production on degraded wastelands and common grazing lands should be taken up as long- term strategy in drought-prone areas.
- Even with the best intention and efforts, it would not be possible to provide relief to all the animals in the affected areas and a large proportion of livestock farmers would migrate to other areas in search of drinking water and grazing. The concerned states should organize migration in planned manner providing required inputs and market support on the migration route. Service centres should be set up on the way and the migrating farmers should be provided security.

CONCLUSIONS

Livestock production continues to be an important occupation for farmer settled in drought-prone areas of India. Therefore, strategies to mitigate the impacts of drought on livestock require multi-institutional and multi-disciplinary approach with short and long-term measures. Contingency planning for feeding and management of livestock during drought will increase both short-term benefits and long-term sustainability through optimum utilization of available feed and fodder resources. Thus, farmers should use different feed resources rationally and adopt the feeding management based on appropriate nutritional technologies to sustain health and production of livestock during drought.

REFERENCES

Bhasin NR 2002 Drought and livestock, Indian *Dairyman* **54 (8)** 33-35

Holmann F, Rivas L, Urbina N, Rivera B, Giraldo LA, Guzman S, Martinez M, Medina A and Ramirez G 2005 The role of livestock in poverty alleviation: An analysis of Colombia. *Livestock Research for Rural Development* **17 (1)** 1-7

IGFRI 1987 Fodder and feed scarcity during 1987 drought: Action plan, IGFRI, Jhansi, 10 Pages

Misra AK, Subrahmanyam KV, Sankar Babu MV, Reddy TY, Shivarudrappa B and Ramakrishna YS 2006 Improving the Livelihood of Landless and Marginal Farmers through Sheep Rearing in Semi-Arid India. *Livestock Research for Rural Development.* **16(5)** 1-11 http://www.cipav.org.co/lrrd/lrrd18/5/misr18073.htm

Negi DS 2002 Drought and livestock farming, *Indian Dairyman* **54(8)** 29-32

Patil BD, Singh J and Pachauri VC 1979 Fodder and forage under drought conditions. IGFRI, Jhansi, 28 Pages

Rangnekar D V 2006 *Livestock and livelihoods of the underprivileged communities in India: review.* International livestock research Institute, Nairobi, Kenya, 72 page

Ranjhan SK and Khera RC 1976 Feeding farm animals during scarcity, ICAR, New Delhi 34 Pages

Shukla PC, Talpada PM and Desai MC 1988 Scarcity feeds and feeding of animals during droughts, *Indian Journal of Animal Production and Management* **4** 184-196

Singh G 2002 Drought management strategies: agroforestry and livestock issues. In: Report of the FAO-CRIDA expert group consultation on farming systems and best practices for drought-prone areas of Asia and the Pacific Region, CRIDA, Hyderabad, 136-152

Singh HP 2002 Farming systems and best practices for drought-prone areas in India. In: Report of the FAO-CRIDA expert group consultation on farming systems and best practices for drought-prone areas of Asia and the Pacific Region, CRIDA, Hyderabad, 84-104

Singh K and Prasad CS 2002 Potential of nutritional technologies in improving livestock productivity. In: Brothal, P. and Parthasarthy Rao, P (eds). Technology options for sustainable livestock production in India. Proceedings of the workshop on documentation, adoption and impact of livestock technologies in India, 18-19 Jan 2001, ICRISAT, Patancheru, India, 132-146

Singh K and Schiere JB 1995 Handbook for straw feeding systems: Principles and applications with emphasis on Indian livestock production. Indo-Dutch Project on Bioconversion of Crop Residues ICAR, New Delhi

World Bank. 1999. *India: Livestock sector review: Enhancing growth and development.* The World Bank and Allied Publishers: New Delhi.

21

Quality Planting Material Production Options and Potential towards Drought Mitigation

P. Ratha Krishnan

INTRODUCTION

Mitigation is the first stage of the four emergency management phases which is followed by preparedness, response and recovery from disaster. It aims at minimizing future threats posed by disasters including drought. Drought is a situation which is dynamic in nature and may occur at any time. In general drought management mechanisms like contingency planning are important to sustain the agriculture production. In particular, trees plays major role in drought management in arid region of India. The trees/shrubs on any land have been referred as a vegetation type (Champion and Seth 1968) however, they are presently recognized as agroforestry system under trees outside forest/tree cover category by Forest Survey of India. Rural folk in Thar Desert have been practicing arable cropping in association with these scattered trees on crop fields since time immemorial. This combined productive, protective system of integrating trees into farming system is rooted in the principles of ecology, productivity, economics and sustainability. In fragile ecosystems like Thar Desert, agriculture alone cannot be a dependable enterprise, hence desert dwellers with their traditional wisdom are integrating trees into farming since ages in order to confer stability and generate assured income (Rathore 2004). As most of the trees are drought resistant, these are able to provide fuel, fodder and other products, when arable crops fail due to drought and other common features of the hot arid region.

Among the perennial trees, Khejri (*Prosospis cineraraia*), Kumat (*Acacia senegal*) and Marwar teak (*Tecomella undulata*) are well accepted agroforestry

Central Arid Zone Research Institute, Jodhpur 242003 (Rajasthan), India
Email: raha_forestry@yahoo.co.in
(Presently at Krishi Vigyan Kendra, Indian Institute of Spice Research, Kozhikode)

species of arid region. Israeli babool (*Acacia tortilis*) and Vilayathi babool *(Prosopis juliflora)* are important for sand dune stabilization and wood purposes. Likewise, ber varieties viz.. Gola. Seb. Kaithli, Mundia, Umaran, Banarasi Karaka and Goma Kirti, and pomegranate, aonla varieties are some important horticulture species of this region. Other than growing the above trees as boundary plantations, practicing integrated farming system and organic farming are also important accepted strategies towards drought mitigation in arid region.

ROLE OF TREES IN DROUGHT MITIGATION

The stand of trees creates a microclimate within the canopy of trees which is cooler, wetter, and has altered soil chemistry compared to the area outside. The altered climate found within forests can support organisms that cannot survive outside the tree canopy. Cooling of air temperatures due to the effect of trees has been well documented in the past through various studies. A tree can be regarded as a natural "Evaporative cooler" using up to 100 gallons of water a day (Kramer and Kozlowski 1960). The release of moisture into the atmosphere through the stomata of leaves is caused due to the evaporation of water at the stomata into the air, thus causing cooling of the air. Hence, cultivation of trees has tremendous potential to modify the harsh climate of arid region to a desirable microclimatic area. Trees on farm bring about favorable changes in the microclimatic conditions by influencing radiation flux, air temperature, wind speed, saturation deficit of understory crops, all of which will have a significant impact on modifying the rate and duration of photosynthesis and subsequent plant growth, transpiration, and soil water use (Monteith et al 1991). A decline of 0.1°C to 0.7°C was evident beneath the canopy cover of *A. tortilis* than that of surrounding open areas at 07 hrs. Around 14 hrs, this decline was of the order of 0.6°C to 2.0°C. Similarly, it was found during monsoon season that the soil temperature just beneath the tree cover was lower by as much as 10°C to 16°C in top soil zone and 4°C to 5°C at 30 cm depth when compared to open field conditions, thereby indicating better soil-thermal regime. Seven year old *A. tortilis* trees provided more humidity on cluster bean grown with the trees. This proves the efficiency of agroforestry practices for maintaining favourable moisture status on arable lands (Ramakrishna and Shastri 1977). Hence, cultivation of trees will reduce the impact of drought on the associated crops and animals. For this purpose, trees need to be prepared in a nursery.

NURSERY AND ITS IMPORTANCE

A nursery is a managed site, designed to produce seedlings grown under favorable conditions until they are ready for planting. All nurseries primarily aim to produce sufficient quantities of high quality seedlings to satisfy the needs of users. Near planting site availability of nursery and supply of quality seedlings for ensured and good survival leads to the reduction of transportation cost and avoids jeopardizing of plant supply during wet season (Evans 1982) which also ensures the success of plantation drives (Ratha Krishnan 2008). The success or failure of plantations in tree planting activity can be forecast by inspecting the nursery, composition of the planting stock and the health of seedlings in nursery (Luna 1989). The success of nursery depends on the operations such as: selection and development of suitable site, practicing appropriate cultural methods, timely execution of operations, protection from pest, diseases and other damages, planning, forecasting, efficient supervision and administration (Ratha Krishnan 2009).

Generally nurseries are classified based on the availability of seedling and its permanency. It includes a) Permanent nursery, b) Flying / temporary nursery, c) Extension nursery. Based on the ownership, the nursery can be classified as departmental nursery, private nursery and corporation nursery. Based on the species composition it can also referred as single species nursery (Example: Casuarina nursery, Teak nursery, Eucalyptus nursery, etc) and mixed species nursery. Seedling nursery and clonal nursery are the types based on the mode of propagation followed in that nursery.

CALCULATION OF NURSERY SEEDLING PRODUCTION AND AREA REQUIREMENT FOR NURSERY

Time of sowing / initiation of propagules production depend on how long the seedlings will take to have an optimum size of a seedling with good rooting and about 20cm tall and coincidence of its ready availability at the time of initiation of monsoon (July for South West monsoon and October for North East monsoon region).

The number of plants required to be produced from a nursery can be calculated as below. Number of plants required for the season = W

Mortality in nursery = X

Transportation / culling loss = Y

Seedling required of buffer loss= Z

Total seedlings required to be produced from the nursery = W+X + Y +Z.

The area of nursery should be about 0.4 ha (1 acre) for every 10,000 seedlings. It also requires perennial or daily supply of water @ 200 litre per 1000 seedlings.

REQUIREMENT OF A GOOD NURSERY

Near planting site availability of nursery and supply of quality seedlings are the basic requirements to reduce the cost of seedlings. Besides, nursery site should be located in the nutrient rich or medium soil, near to water source, free from soil pathogens and insects, availability of cheap and skilled labors and has good access to the main road for easy transportation. Designing and layout of nursery should be carried out by considering the above along with the availability of resources. Generally a good nursery should consist of water tank /pond, water pump / pump house, seed & fertilizer store room, implement shed, germination/ mother bed area, potting / container filling area, seedling raising area, worker mess, office room, etc. Initially the infant seedlings require protection from direct sun and hot winds. It can be provided to the seedlings by maintaining them in a partial shade condition by planting of trees in the nursery or by the artificial shade houses / screens.

MANAGEMENT OF NURSERY

The success of nursery depends on the operations such as: selection and development of a suitable site, use of appropriate cultural methods, orderly timing of operations, protection from pest, diseases and other damages, planning, forecasting, efficient supervision and administration The main operations in nursery include:

1. Planning, controlling and recording all stages from receipt of seed to sale of seedlings.
2. Seed collection, purification, storage, pretreatment, etc.
3. Soil preparation and making of mother bed and filling of container.
4. Basal fertilizer management and top dressing.
5. Sowing of seed or placing of cutting in the bed and its transplanting to the containers.
6. Operations such as picking out, watering, weeding, shifting, etc.
7. Protection against human, animals, fungi, insects, nematodes, etc.
8. Packing and dispatch of plants.

MANAGEMENT OF PEST AND DISEASES

Pest control is necessary whenever symptoms of attack by viruses, bacteria, fungi and insects were noticed in the nursery. The pathogens are either carried in the seed, soil borne or air transmitted. Pretreatment of seeds, sterilization of soil and foliar spray of chemicals are the recommended methods to control them. The fungicides used for this are generally composed of Copper (ex. Bordeaux mixture, Perenox, Cuprosan, Zincop, etc), Sulphur (ex. Lime - Sulphur, Organo-Sulphur, etc.) or Organo- mercurials (ex. Agrosans, Ceresan, Sem esan, etc.). Insects attack should be prevented by maintaining hygiene in the site, by keeping weeds in check and by maintain the health of seedlings in optimum condition. The insecticide may be of a) Inorganic (Arsenicals, Fluorsilicates) b) Organic (Plant origin- Pyrethrins, Rotenones, Ryania; Synthetic - Aldrin, Aldrex; Chlorinated hydrocarbons - BHC, DDT, Endrin, Endosulfan; Organophosphates-Parathion, Malathion, Dimethoate, Monocrotophos; Caramates - Mimetilan, Carbofuran, etc.).

GENERAL INFORMATION ABOUT PROPAGATION OF TREE SEEDLINGS

Due to the long gestation period and unavailability of improved variety seeds of trees, the production of genetically improved seedling has certain limitations. The general practice to overcome this limitation is to collect fresh and viable seeds from the good mother tree and use it for propagation. In case of short rotation species such as *Casuarina*, Eucalypts, and *Leucaena* the improved clones are available and they have to be used as mother tree for further propagules production. So, it is clear that the long rotation species are mostly propagated sexually whereas the short rotation species, asexually. Sexual propagation means propagation of species through seeds, which expresses segregation. Asexual propagation means the propagation of species by cuttings, grafting, budding, air-layering or by micro- propagation, which gives true to type progenies. Each method has its own advantages and limitations. True to type propagules, early flowering and maturity, improved quality and all season production and availability of planting material are the advantages of clonal propagation. Requirement of special propagation structures, skilled labor, higher initial costs are its limitations.

Sexual propagation

The main requirement of this method is the fresh, viable and quality seeds. Mother bed (raised or sunken), containers (Polybags / pots / dona / root trainers), nursery mixture (Forest soil, sand, FYM @ 2: I: 1 ratio) are the other requirements. Trees generally propagated by seeds, and its number of seeds/Kg, best seed collection period, viability, pre-sowing methods to break dormancy and germination percentage are set in Table 1.

Table 1 Details of seedlings preparation of some tree species

Species	No. of seeds per kg	Best seed collection time	Pre-sowing treatments	Viability	Germination %
Acacia nilotica	7000-11000	April-June	Fresh, 40-50 min H_2S0_4 scarification	>2 yrs	88
Ailanthus excelsa	9500	March-June	Dewinking, Watersoaking 12-24hrs	6months	70
Albizia lebbeck	9200	Nov-Dec-Feb	Water soaking for 24hrs	4-5 yrs	40 - 60
Azadirachta indica	1750-4000	June-July	Fresh, Depulping and sowing	2 weeks	50
Casuarina equisetifolia	7,60,000	June-Dec	Fresh, Mixed with ash & BHC	12 months	70
Eucalyptus hybrid	>3,57,000	Sep-Dec	Fresh, apply with ash/ sand & BHC	2 yrs	90
Hardwickia binata	3900	April-May	Fresh, No treatment	1 yr	60-80
Jatropha curcas	1200–1800	Feb–Mar, Oct-Nov	Fresh, Water soaking 12–24 hrs	1 yr	>90

Leucaena leucocephala	NA	April	Hot water treat for 2-3minutes	2 yrs	90
Pongamia pinnata	800-1500	March-May	Fresh, No treatment	months	>80
Prosopis juliflora	12500	May-June	20 min H_2SO_4 scarification	18 months	80-90
Simarouba glauca	1000-1500	Jan-April	Fresh, No treatment	1 yr	>80
Syzygium cumini	1200	June-August	Fresh, No treatment	1 month	50
Tamarindus indica	1000-1200	Oct-Dec	Fresh, 15min. H_2SO_4 scarification	3 yrs	>60
Tectona grandis	2500-3000	March-April	Alternate wetting & drying for 7 days	<2 yrs	30-60

(Banerjee 1987, Ratha Krishnan et al 2008)

Asexual propagation

It is by means of cuttings, grafting, layering and micro-propagation. It requires the explants from the improved / selected mother tree. The required propagation structures for this method are: green house, mist chamber with cooling pad, polytunnels, shade / screen house, tissue culture room with temperature and light control. The required hardwares are root trainers (Composite / block / single cell), other containers, grafting machine, secateur, cutting knife, scissors and vermiculite. Details of proven vegetative propagation methods of some multipurpose tree species are given in Table 2.

Table 2. Proven vegetative propagation methods for the MPTs.

Species	Type of propagation	Best month	Treatment	Reference
Acacia mangium	Semi hardwood cutting	May- June	IBA	(Bulgannawar, 1989)
Azadirachta indica	Cutting, grafting, air layering	Feb- March	IBA	(Ratha Krislnan et al 2005)
Casuarina Spp	Softwood I nodal cutting	All months	IBA	(Lundquist & Torrey 1984)
Eucalyptus camaldulensis	Softwood cutting	Aug- Sep	Auxins	(Rinallo et al 1987)
Eucalyptus hybrid	Coppice shoots & leafy cuttings	All months	400 ppm IBA	(Gurumurti et al., 1989)
Hardwickia binata	Stem cutting	Jan- March	4000 ppm IBA	(Handa, 2003)
Pongamia pinnata	Young shoot with heal cutting	April	4000 ppm IBA	(Thomas, 1988)
Populas deltoides	ETP cuttings	All months	Auxin	(Greg Morgenson, 1 991)
Prosopis cineraria	Hardwood cutting	All months	Auxins	(Balboa et al., 1987)
Simarouba glauca	Grafting, air layering	Oct- Jan	4000 ppml BA	(Ratha Krishnan, 2001)

NURSERY TECHNOLOGY FOR IMPORTANT TREE SPECIES

Propagation practices of important Indian trees are already elaborated by Banerjee (1987) and Ram Parkash et al.(1991). Nursery practices of some selected tree species are summarized below:

Timber and industrial wood tree species

Acacia nilotica- Freshly collected seeds are sown in polybags (after pre-sowing treatment) at 1.5cm depth during Feb - March. Shading is necessary to prevent surface drying. Avoid excessive watering in the first month. Seedlings are fit for planting out in July-August of the same year. But, one year old seedlings are good for planting in the field.

Ailanthus excelsa- Fresh seed is sown in well worked raised nursery beds, in drills about 20 cm apart, and lightly covered with finely pulverized soil; depth of sowing being not more than 5mm. Seed being minute is mixed with ash or fine soil to ensure even sowing. Seed beds are irrigated regularly but lightly as the seedlings are liable to damping off. Germination commences in about 10 - 12 days and takes about 50 days to complete. Young seedlings are susceptible to insect and rat damages. In order to avoid it, porcupine proof fencing of nursery is helpful. Seedlings raised in the containers attain 15- 25 cm height in the first growing season, while the subsequent growth is more rapid.

Albizia lebbeck - Seedlings can be raised by sowing in drills 15 cm apart in the nursery beds in March, or in containers. Germination commences in 4 days and is complete in 17 days. Prick out seedlings into polybags when they are 7.5 cm tall. Seedlings growth is fast and is susceptible to insect attack also. Seedlings reach plantable size in July of the same year.

Azadirachta indica- Depulped, fresh, dried seeds are sown in nursery raised bed, in lines 15- 20 cm apart at 1 cm depth. Nursery beds are watered sparingly; weeded, hoed and protected against frost in winter. Germination starts in about a week and continues for 3 weeks. Plantable size seedlings (7- 10 cm tall) are produced in 2- 3 months. But the pricked out seedlings are maintained in the nursery for one year to increase its field performance.

Bamboo spp.- Bamboo reproduce themselves naturally from seed fallen on the ground after sporadic or gregarious flowering, also from the rhizome buds, throwing up culms as annual shoots. Bamboo can be propagated in the nursery by direct sowing of seeds, planting nursery raised container or bare rooted seedlings, and by vegetative means. Most common method of vegetative reproduction is by rhizome or offset planting; other methods are layering, nodal cuttings and culm cuttings.

Dalbergia sissoo- Broken pod segments or clean seeds are sown about 1.25 cm deep in nursery beds, 0.9 m to 1.2 m wide, fiat, or sunken beds during Feb-March. Sowing may be done any time between mid-February and September, but late sowings give lesser growth. Bed needs to be irrigated at least twice in a year. For planting out, 3-4 months old entire plants, when about 60 cm tall are suitable; for making stump 12- 16 months old stocks are preferred. Before digging out plants, most of the lower leaves are stripped off to reduce transpiration.

Dendrocalamus strictus - Seed is broadcast in raised nursery beds, 1m x lm, in September- October; seed rate being approximately 1.5 kg per bed. Germination takes 10 – 20 days; seedlings about 3 months old and 4-5 cm tall are pricked out in polythene bags. Seedlings are fit for planting out during following July; seeds may also be sown direct in polythene bags. Seedlings with 45- 60 cm tall in one year are fit for planting out.

Eucalyptus spp. - For raising container plants seed is first sown in raised nursery beds and then pricked out into polythene bags, after germination. About 200g seed is sown per germination bed of 1 m x 1 m size, during Sep - October or after Feb - March after winter frost. To avoid dense sowing, seed is mixed with fine sand or earth, and given a light covering of fine sand and insecticide @ 100g per bed is advised to protect seeds from insect and termite damage. Germination beds are covered with thatch grass to prevent drying out and minimize damage by birds, or direct impact of water spray or a chance shower of rain. Germination takes place within 5 - 15 days and then thatch is removed. Seedlings are pricked out into polythene bags when about 5- 10cm tall, normally 4-6 weeks after germination. These bags are placed in sunken beds and watered regularly up to its 1 to 1.5 m tall and are fit for planting in 6-8 months. Eucalyptus is also propagated by two leafy cuttings which require mist chamber and other propagation hardware. This ensures true to type propagules and higher productivity.

Gmelina arborea- Seed is sown in drills, as soon as it ripens, in nursery beds; shading is necessary. Germination starts in 2- 3 week. Seedlings are transplanted in polybags and maintained up to 6 months for its height growth of >10 cm.

Tectona grandis- The soil is well dug up, mixed with ashes, in a well-drained site. Standard size of bed is 12 m x 1.2m; raised bed in moister localities to avoid water logging and flat or slightly sunken bed in dry zone are advised. Treated seeds are broadcast or sown 7.7 x 7.5 cm lines, about 1 cm deep, between February and May. Seedlings are pricked out in transplant beds during the first rains and one year old seedlings are suitable for the stump preparation. 1.0 to 2.0 cm collar diameter; 3 -4 months old seedlings are suitable for planting.

Fuel wood tree species

Acacia auriculiformis - Seeds requires pre-treatment, either 24-48 hrs immersion in water at room temperature or boiled water for 24 hrs. Sowing in bed / polythene bag (20 x 15 cm size) during March to early April is suggested. Seed is sprinkled with kerosene oil to protect against ants. Germination starts in 30 - 35 days. A thatch barrier on one side of bed is needed to protect the young seedlings against hot winds. 15 - 30 cm tall seedlings are ready for planting in the fields.

Acacia senegal-Seeds are sown in polythene bags filled in with a mixture of good loamy soil and FYM in the ratio of 2:1 during June-July. One year old entire seedlings are planted out in the following monsoon. Seedling is protected against drought and frost by regular watering.

Casuarina equisetifolia- Nursery soil should be inoculated with soil collected from *Casuarina* plantation, and gammexane sprinkled around beds to prevent ant damage, seed is sown broadcast about 10 g per sq. m, usually in November, and lightly covered. Beds are mulched with palm leaves or straw and watered daily till germination is complete (20-30 days). Seedlings are fit for planting when 35-45 cm tall in 5-6 months.

Pongamia pinnata- Seed is dibbled in the nursery beds any time after collection, preferably in the beginning of hot weather at a spacing of 7.5cm x 15 cm. Mulching of sown beds is helpful. Germination commences after about 10 days and completes in a month. Pricing out improves growth. Seedlings attain a growth of 25-30 cm by the end of first growing season, and of about 60 cm by the beginning of the next rainy season, when they may be planted out entire, with ball of earth, or in the form of stumps.

Prosopis cineraria- Pre-treated seed is sown in soil filled perforated polythene bags; 2 seeds in each, 1 cm deep, in June- July. Regular watering is necessary, but excessive watering is harmful. Seedlings are regularly weeded; if two emerge, one of the two seedlings is removed. They reach planting size after one year.

Prosopis juliflora- One seeded pod segment or seeds are sown in properly worked up soil and lightly covered with soil in March-April or earlier in mid-February. The germinated seedlings are transplanted to polybags up to 2 months. The seedlings attain1.5 m height in one year and are ready for planting in the field.

Fodder tree species

Albizia procera- Treated seeds (hot water or cold water soaking treatment) is sown in nursery beds in drills 7.5 x 7.5 cm in May. Light watering is given for about a week. Germination commences in 3 to 4 days and completes in 3 weeks. Seedlings attain 13- 15 cm height in July of the same year, and are fit for planting out, with or without ball of earth, during rains.

Leucaena leucocephala- Seeds require inoculation with *Rhizobium* culture to ensure nodulation and nitrogen fixation. Plants may be raised in nursery beds or containers. Pre-treated and preferably inoculated seeds are sown in beds in March-April in lines spaced 30 cm apart and seeds 5 cm apart in lines and about 1.2 to 1.5 cm deep. Nursery soil should be neutral to mildly alkaline. Beds are irrigated regularly but lightly. Germination commences in 4 - 6 days and the seedlings attain plantable size by July - August.

Moringa oleifera- Seed is sown in June, in sunken nursery beds, in lines 20 cm apart, seeds about 2 cm apart in lines and just about 1 cm deep. Addition of well rotten compost to the nursery bed is beneficial. The required seed rate is 30-35 g/ sq.m of bed. Germination starts in about 8 - 10 days and takes 4 weeks to complete. The seedlings are fit for planting out, and also for making stumps after one year.

Sesbania grandiflora - Seeds are sown in polythene bags after collection in May, and watered regularly. Germination starts on the third day of sowing and is complete within a week to ten days. Seedlings grow very rapidly and are fit for planting out in about one month.

Ziziphus mauritiana- Treated seed are sown broadcast, or in lines in nursery beds, or directly into containers, in February- March, and covered with soil to a depth of 1 .5 to 2.0 cm. Seedlings are pricked out into transplanted beds / containers at the 4-leaf stage, at a spacing of 30 x 15 cm. Plants attains plantable size (>30 cm) in the first or second rains. Cultivated varieties are usually propagated by ring grafting or budding on wild root-stocks, either *in-situ* or on seedlings at nursery stage.

CONCLUSION

Drought and other adverse climatic effect can be minimized by using the tree cover. That is why maintaining one third of our geographic land under green cover is recommended by forest policies. The success and productivity of perennials mainly depend on the quality of inputs especially the seedlings used for planting. Selection of right species and varieties are important to overcome adverse conditions. Using quality planting material and managing it properly has much more potential to develop green belts. Development of green belts will improve the microclimate. Hence, production of quality seedlings in nursery is a way for minimizing adverse impacts especially of droughts.

REFERENCES

Balboa O, Cotes OI and Arce JP 1987 *Interciencia* **12 (1)** 27-31

Banerjee, AK 1987 *Nursery manual for species commonly used in social forestry in India*, Rep: National Wasteland Development Board, Kapoor Press. New Delhi, 64p.

Barbara, ET 1985 Seedlings morphological evaluation - What you can tell by looking. In Proceeding: Evaluating seedling quality: Principles, procedures and predictive abilities of major tests (Eds. Duryea), Workshop held at Forest Research Laboratory, Oregon State University, Corvallis, pp. 59- 71

Bulgannawar, GN 1989 Seminar on vegetative propagation. ICFRI : IFGTB, Coimbatore.

Champion, HG and Seth, SK 1968 A Revised Survey of Forest Types of India, Govt. of India Press, New Delhi, 404 p.

Evans, J 1982 *Plantation forestry in the tropics* Oxford University Press, Oxford. 471 p.

Greg Morgenson, 1991. Vegetative propagation of Poplar & Willow, Intermountain Forest Nursery Association Annual Meeting, Park City, Utah, August 12-16, P. 3.

Gurumurti, K Dhawan Mcena and Bhandari, HCS 1989 *Vegetative propagation*. ICFRE, Institute of Forest Genetics and Tree Breeding, Coimbatore.

Handa, AK 2003 Final report on "Eco- physiology of regeneration in MPTS of Semi-arid Tropics" submitted to ICAR New Delhi, P. 67.

Kramer, PJ and Kozlowski, TT 1960 *Physiology of trees*. McGraw-Hill publications, Mc Graw Hill, 642 p.

Luna, RK 1989 *Plantation Forestry in India*. Publishers International Book Distributors, Dehradun, India, pp.93-122.

Lundquist, R and Toney, JG 1984 *Bot. Gaz.* **145(3)** 378-384.

Mohammad, G 1993 Application of biofertilizers in production of heal thy and vigorous transplants in forest nursery. In: Nursery technology for agroforestry, Eds. Sunil Puri and P.K. Khosla. Oxford & IBH publishing Co. Pvt. Ltd, New Delhi, pp2 47 -262.

Monteith, JL Ong, CK and Corlett, JE 1991 Microclimate inter-actions in agroforestry systems. Forest Ecology Management, **45** 31–44.

Rama Krishnan and Shastri, 1977 Arid zone studies, ENVIS, p 4.

Ram Prakash, Chaudhari, DC and Negi, SS 1991. Propagation practices of important Indian Trees. IBD Publishers, Dehradun, 452 p.

Ratha Krishnan, P 2001 Studies on elite seedling production in *Simarouba glauca* (Linn.) by sexual and asexual propagation. M.Sc (Forestry) thesis, Tamil Nadu Agricultural University, Coimbatore, 141 p .

Ratha Krishnan, P 2008. Nursery management and quality plant propagation. In: Training manual on "Advances in Agroforestry Research and Development" held during 24thNovember- 11 December, 2008 at NRCAF, Jhansi, pp.85-91.

Ratha Krishnan, P, Gupta, VK and Dhyani, SK 2008 Two decades of Academic Achievements of N RCAF (1988 -2008). Published by NRCAF, Jhansi, P 86.

Ratha Krishnan, P, Paramathma, M and Parthiban, KT 2005 Screening and multiplication of superior genotypes of neem for quality wood production. Proceedings of IUFRO International Conference on multipurpose trees in the tropics: Assessment, growth, and management, Jodhpur, Nov 22 -25, pp. 679-684

Ratha Krishnan, P, Sundersingh Rajapandian, J and Paramathma M 2008 Standardization of nursery technology for quality *Simarouba glauca* seedling production. In: Exotics in Indian Forestry. Eds: Sanjeev K. Chauhan. Etal., Pub: Agrotech Publishing Academy, Udaipur, pp.646- 650.

Ratha Krishnan 2009. Nursery techniques for timber, fuel and fodder species for watershed development. Training course compendium: Watershed management through alternate land use system, NRCAF, Jhansi, p 78 – 84.

Rathore, NS 2004 Fauna of desert national park conservation area series, 19 of Zoological Survey of India Publications, Kolkata, India, p. 135.

Rinallo, Gellini, R and Fabbri, A 1987 Advances in horticultural Sciences, 1 (I): 27-23.

Thomas, KI 1988 FRC Pub. (43) vii+ 52 pp. 17 ref. Samdakan.Sabah, For. Res. Centre.

22

Wind Erosion Hazards during Drought and its Mitigation

Priyabrata Santra, Suresh Kumar and S. Poonia

INTRODUCTION

The arid western Rajasthan is prone to frequent occurrence of drought due to low and erratic nature of rainfall. Detailed analysis on drought in arid Rajasthan during the period 1901-2012 revealed the occurrence of drought once in three to four years (Ramana Rao *et al*., 1981). Jaisalmer district in western Rajasthan is most prone to drought, which occurs almost in alternate years. During 1901-2012, the agricultural drought in Jaisalmer occurred in 70% of the years, out of which drought in 43% years was of severe in nature and in 27% years moderate, thus drought affect considerably the crop and fodder production. Next to Jaisalmer district, the Barmer district experienced severe drought in 30% years and moderate drought in 20% years. Bikaner district experienced severe agricultural drought in 23% years and moderate in 25% years, whereas, Jodhpur district experienced severe drought in 17% years and moderate drought in 28% years. Interestingly, it has also been observed that western districts of Rajasthan which are mostly affected by droughts are also affected by wind erosion. Frequency of droughts in different districts of western Rajasthan is presented in Table 1.

The worst drought conditions were experienced during the years 1918 and 2002. Prolonged drought put tremendous stress on natural resources of the region, which leads to severe scarcity of food, fodder and water (Rao 1997; Rao et al., 1997 and Rao and Vijendra, 2005).

The occurrence of drought is mostly associated with low occurrence of rainfall, high temperature and strong wind speed. A typical trend of annual rainfall pattern at Jaisalmer given in Fig. 1 depicts the picture clearly. Once in

ICAR-Central Arid Zone Research Institute, Jodhpur, Rajasthan 342003
Email: priyabrata.iitkgp@gmail.com

every three to four years, annual rainfall was very low than normal however in succeeding year it was high, following a sinusoidal pattern. For example during 2002, which was a severe drought, the annual rainfall was only 28.5 mm and same was the case during 1987 also. It was an alternate year drought during 2004 but in 2005-2007, annual rainfall was more than average (300-400 mm).

Table 1. Frequency of different intensities of meteorological droughts in arid Rajasthan

District	Period	Mild	Moderate	Severe	Total years	% years
Ganganagar	1926-2012	20	13	15	48	57
Churu	1906-2012	34	17	05	58	55
Hanumangarh	1906-2012	17	16	19	52	49
Jalore	1901-2012	31	17	16	64	58
Nagaur	1901-2012	29	23	14	66	60
Jhunjhunu	1901-2012	28	21	09	58	53
Jodhpur	1901-2012	24	23	16	63	57
Sikar	1901-2012	23	21	15	59	54
Barmer	1901-2012	22	18	22	62	56
Bikaner	1901-2012	22	21	13	56	51
Pali	1901-2012	18	30	11	59	54
Jaisalmer	1901-2012	15	23	23	61	55

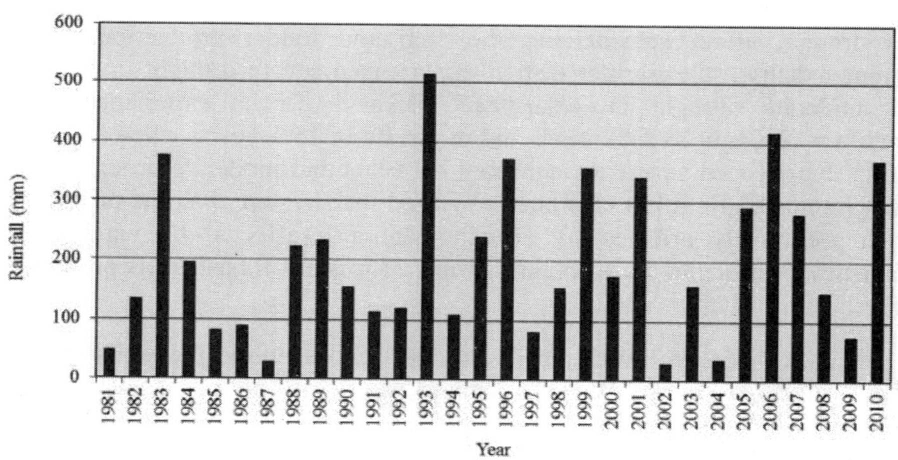

Fig. 1. Annual rainfall pattern in Jaisalmer during 1981 to 2010.

WIND EROSION AND DROUGHT

Severity of wind erosion and occurrence of drought are very much related to each other and both are governed by prevailing weather condition. Drought is associated with very low rainfall, high evapotranspiration and strong wind which do not support vegetation cover on ground. And if such situation prevails, wind erosion process turns very severe because drought makes the land surface favorable for this natural land degradation process. Wind erosion occurs mainly through strong wind forces on loose sandy soils and thus mostly affected by wind erosivity and soil erodibility. If both these factors are very high, the severity of the wind erosion process will be high. In the following section we briefly describe how different causative factors of wind erosion are affected by drought. We also discuss a few experimental results from field situation at Jaisalmer.

Factors affecting wind erosion

Wind erosion process is affected by wind erosivity, soil erodibility and surface conditions. To represent these main three causative factors, different subfactors have been identified which can be quantified in field for evaluating the impact of wind erosion. Most primitive wind erosion evaluation system was developed by Woodruff and Siddoway (1965), which is known as wind erosion equation (WEQ) and expressed as follows:

$$E = f(I, C, K, L, V) \quad (1)$$

Where E is the potential annual soil loss (t ha^{-1}), I is soil erodibility factor, C is climatic factor, K is soil surface roughness factor, L is equivalent width of field (the maximum unsheltered distance across the field along the prevailing wind erosion direction), V is equivalent quantity of vegetative factor. Later on, this empirical relationship was further modified to revised wind erosion equation (RWEQ) by incorporating few process based inputs to improve the accuracy of estimation (Fryrear et al., 2000). In RWEQ, each factor was tried to be quantitatively described e.g. weather factor, soil erodible fraction factor (EF), soil crust factor (SCF), soil roughness factor (K), combined crop factor (COG) including flat residue factor (SLR_f), standing residue factor (SLR_s), crop canopy factor (SLR_c) etc. Among these factors, which are directly affected by drought are discussed below:

Weather factor

Weather factor (*WF*) in RWEQ is calculated with an aim to estimate soil loss through RWEQ approach. Wind factor (*Wf*) and soil wetness (*SW*) may be calculated separately to obtain *WF* as follows:

$$WF = W_f \times SW \times SD \quad (2)$$

Where, W_f is the wind factor, *SW* is the soil wetness factor and SD in snow depth factor. The W_f and *SW* factor are very important factor since the first one directly calculates the kinetic energy of wind whereas the second factor defines the surface dryness quantifying how much the soil is prone to erosion. The *SD* factor may be ignored for hot arid situation.

In RWEQ, the basic driving force is wind, which is calculated as follows:

$$W_f = \sum_{i=1}^{N} \rho \frac{(U_i - U_t)^2 U_i}{gN} \tag{3}$$

where W_f is the wind factor (kg m^{-1} s^{-1}), U_i is the wind speed at 2 m height (m s^{-1}), U_t is the threshold wind speed, N is the number of wind speed observations in a time period, ρ is the air density (1.293 kg m^{-3}), g is the acceleration due to gravity (9.8 m s^{-2}). It is clear from the Eq (3) that a small increase in wind speed will have a large impact on wind factor since wind factor is proportional to 3rd power of wind speed. For example, increase in wind speed from 6 m s^{-1} to 7 m s^{-1} will increase the wind factor by 4.6 times if we consider U_t as 5 m s^{-1}, which was reported by Fryrear et al. (1998) as the threshold wind speed at 2 m height from ground surface. However, U_t depends on ground surface conditions which vary from place to place. Even, in western Rajasthan, wind speed has been measured at 3 m or 10 feet height from ground surface at most meteorological stations. In India, reports by Gupta et al. (1981) and Ramakrishna et al. (1990) revealed the threshold daily average wind speed for Chandan, Bikaner and Shergarh (Jodhpur) as 2.78, 1.39 and 1.11 m s^{-1}, respectively at 3 m height. Averaging and conversion of these reported threshold values from 3 m height resulted in to the threshold value at 2 m height as 1.66 m s^{-1}. Wind factor at Jaisalmer considering U_t as 1.66 m s^{-1} during a severe drought year (2002) and normal year (2007) are presented in Fig. 2. Detailed description of wind factor for Jodhpur may also be found in Santra et al. (2013). It is clearly visible from the figure that wind factor was quite high during hot summer months during a drought year than normal year.

Soil wetness factor (*SW*) in RWEQ is calculated as follows:

$$SW = \frac{ET_p - (R+I)\dfrac{R_d}{N_d}}{ET_p} \tag{4}$$

where, ET_p is potential relative evapotranspiration R_d is number of rainfall and/or irrigation days, R+I is rainfall and irrigation (mm), N_d is the number of days (normally 15).

$$ET_p = 0.0162 \left(\frac{SR}{58.5}\right)(DT + 17.8)$$

where SR is the total solar radiation (cal cm^{-2}) for the total time period and DT is average temperature (°C). From Eq (4), it may be observed that in case of no occurrence of rainfall or irrigation, the SW factor will be 1. If the rainfall and irrigation amount is almost equal of evapotranspiration, *SW* factor will be very close to zero, which indicates the total check in wind erosion. Principally, soil moisture creates fragile soil aggregates and does not allow the soil to blow irrespective of strong wind speed.

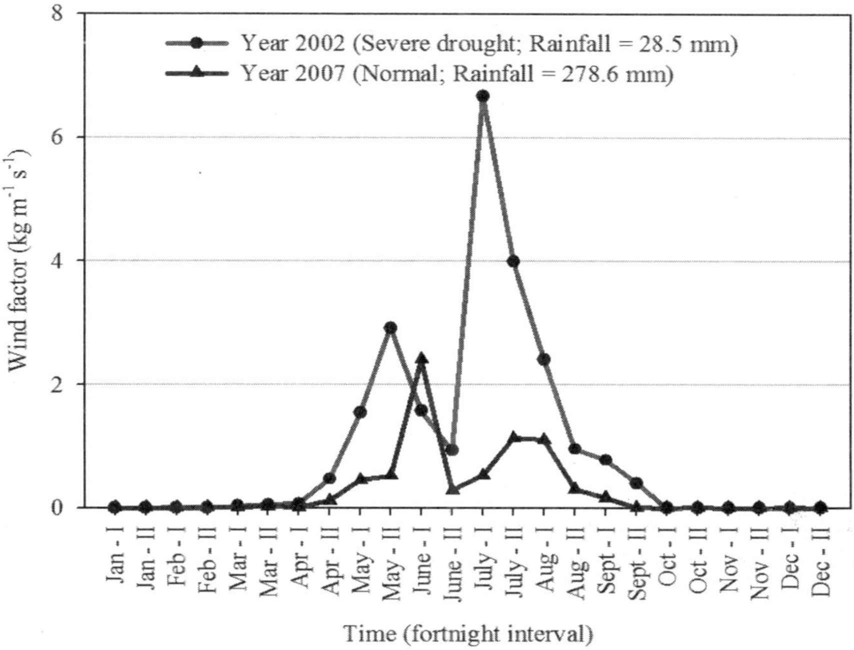

Fig. 2. Wind factor during a severely affected drought year (2002) and normal year (2007) at Jaisalmer

Therefore, description of soil moisture status is necessary for prediction of wind erosion events for either a small region or for a large scale. For prediction of wind erosion events at a small scale, frequent observations on soil moisture condition is sufficient but for a large scale it is necessary to model the soil moisture change. An attempt had been made to quantify surface as well as subsurface moisture conditions at wind erosion experimental site of Jaisalmer, Rajasthan during a time period of two weeks from 29th April to 10th May, 2005 when three rainfall events of 2 mm, 25 mm and 7 mm occurred respectively on 29th April, 1st May and 3rd May. Before rainfall events, soil moisture for surface (0-10 cm) was observed as low as 0.40% (w/w), which was improved to 3.96% (w/w) after three rainfall events of 34 mm (Fig. 3). Significant amount of soil moisture was observed for 10-20 cm soil (2.87%, w/w) although soil surface has dried to 1.88% (w/w) after 7 days from last rainfall events of 7 mm on 3rd May.

Soil roughness

Presence of non-erodible clods over undulating soil surface of the desert provides roughness and increases threshold friction velocity. Such type of roughness can broadly be classified into two main categories-random roughness and ridge roughness. Measurement of random roughness in field was difficult and cumbersome and can be done either using a pin meter (Wagner and Yu, 1991,

Skidmore *et al.*, 1994) or through chain method (Saleh, 1993, 1997). However, with the application of laser technology, it is now easy to quantify soil roughness. Ridge roughness is generally computed from ridge height and ridge spacing and their alignment with wind direction (Fryrear *et al.*, 2001).

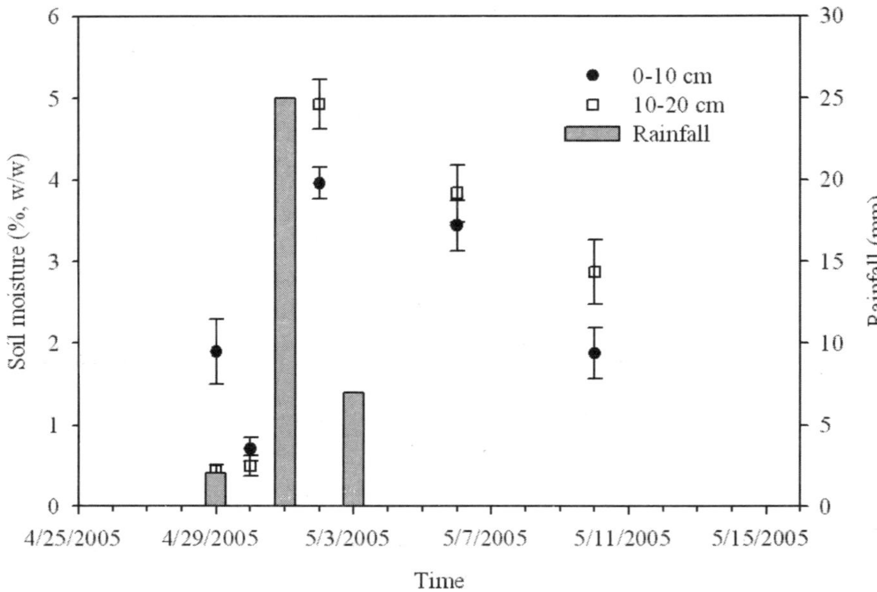

Fig. 3. Soil moisture regime as affected by rainfall events at Jaisalmer

Flat residue factor (SLR_f)

Vegetation covers either in the form of flat residue or standing biomass are other important roughness elements, present over desert area. Flat residue lying over ground surface can be quantified through collecting and air-drying residue content (g m^{-2}) and the related factor can be quantified as follows:

$$SLR_f = e^{-0.0438(SC)} \tag{5}$$

Where SLR_f is the soil loss ratio coefficient for flat cover and SC is the soil surface covered with flat residue in %. For example, if 25% of the surface is covered by flat cover the SLR_f factor will be 0.33. It indicates that if the surface is covered by 25% only, it can reduce the erosion by 2/3rd of the erosion on a completely denuded surface. Kumar et al (2015) have demonstrated that drought results in poor regeneration and low growth of many annuals and even perennials as evident from their less cover assessed by remote sensing and validated by ground samples, for example, in Shergarh dune landscape. Such drastic declines in the amount of available flat residue will results in low SLR_f and enhanced erosion.

Standing residue factor (SLR_s)

Protruding roughness provided by left over stubbles or grass tussocks can be computed through soil loss ratio factor as follows:

$$SLRs = e^{-0.0344(SA^{0.6413})} \tag{6}$$

where SLR_s = soil loss ratio for standing plant silhouette, SA = silhouette area computed by multiplying the number of standing tussocks per m^2 area times average diameter (cm) times tussock height (cm). During drought year the standing residue or silhouette area will be low and hence higher SLR_s are expected. The basic principle of low $SLRs$ factor with higher silhouette area is the change in wind velocity profile over a dense standing residue which in fact increases the threshold wind velocity to cause erosion. In extreme arid tracts of the Indian Thar desert, grazing lands even occupy 85-90% of the total area and thus protects the soil surface from erosive effect of wind too some extent. Moreover, the growth of vegetation in the rangelands starts with the onset of monsoon in the end of June or beginning of July, each year. The top growth is attained in September or October and then it declines upon maturity as most of ephemerals have very short life cycles. However, during a drought year, the ground coverage of annuals is absent due to its failure to germinate while perennial rangeland grasses regenerate very poorly ending up in much less standing residue (Kumar et al., 2015) and higher SLR_s and that makes wind erosion process very severe.

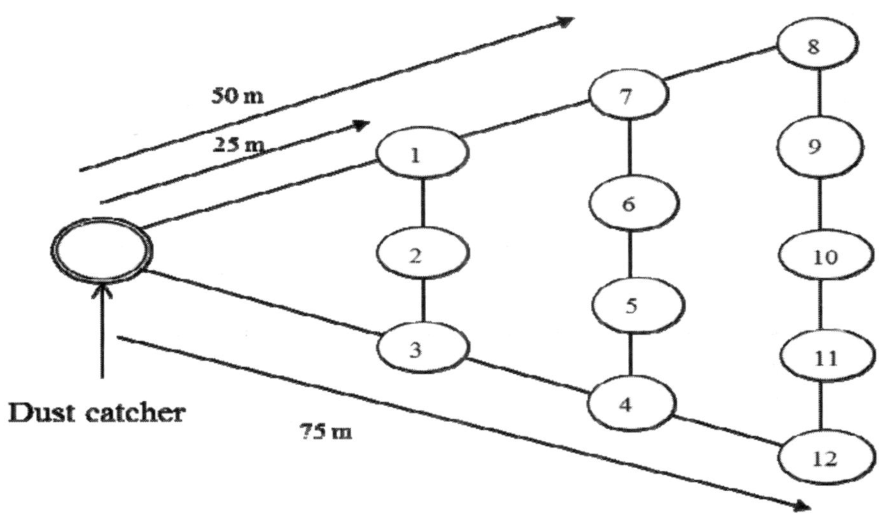

Fig. 4. Diagram of the quadrate studied to quantify standing tussock of sewan grass at Chandan

In a field experiment at Chandan, surface roughness provided by rangeland grasses was directly quantified through measurement of dry grass tussocks of sewan (*Lasiurus indicus*) using quadrate technique using a quadrate of 3 m radius. Schematic diagram of the quadrate design and the measured roughness parameters are mentioned in Figure 4 and Table 2.

Table 2. Surface roughness provided by tussocks of sewan grass

Quadrate No	Number of tussocks	Mean Height (cm)	Mean Diameter(cm)	Mean Distance (cm)
1	18	41.5	55.13	-
2	9	29.75	39	135.5
3	11	29	40.5	195
4	8	32.5	55	184.5
5	19	28.75	32.5	173.33
6	15	25	41.88	206.67
7	9	23.25	51.5	260
8	10	27.75	44.75	233.33
9	7	42.5	51.38	201.67
10	10	29.25	41.88	208.33
11	11	22.75	47	166.67
12	1	34	15	-
Mean	12	30.5	42.96	196.5

During summer months, number of dry tussocks of sewan grass, which act as a silhouette was found 0.4 per m^2. Average diameter and height of sewan tussock was 30 and 43 cm, respectively. Accordingly, SLR_s was calculated as 0.15. However, after rainfall events, dry tussocks immediately sprout and green foliage growth start, which results in higher biomass and low SLR_s value. Generally, after rainfall events after hot summer in desert, wind becomes less erosive and hence the effect of grass coverage was not further quantified during post-monsoon periods. In the pre-monsoon summer months, the SLR_s value of 0.15 contributed by dry tussocks of sewangrass means that during hot summer months even this dry grass coverage can prevent wind erosion by 85% as compared to completely bare soil surface. It indicates that sewan grass coverage even though dry in summer provides such a unique roughness to the sandy surface in controlling wind erosion in desert even during drought period. As compared to the controlled grazing land as mentioned in Table 2, the SLR_s value provided by sewan grass coverage at an overgrazed site at Jaisalmer was found higher due to low tussock density and low height of grass, which resulted in 3-5 times higher aeolian mass transport rate (Mertia et al. 2010).

However, surface roughness can also be indirectly measured through studying the wind velocity profile, which is generally expressed as follows:

$$U(z)=\sqrt{\left(\frac{\tau}{\rho}\right)}.\left(\frac{1}{k}\right).\ln\left(\frac{z}{z_0}\right)=\left(\frac{u_*}{k}\right).\ln\left(\frac{z}{z_0}\right) \quad (7)$$

Where, $\underline{u}(z)$ = wind speed at a height z,

τ = shearing stress (Pa)

ρ = density of air (kg m^{-3})

u_* = Friction velocity (m s^{-1})

k = van karman constant (0.4)

z_0 = aerodynamic roughness parameter (m)

From Eq. (7), it has been observed that wind velocity profile can be characterized by two parameters, u_* and z_0. Aerodynamic roughness parameter (z_0) corresponds to the theoretical height at which wind speed near the surface falls to zero and depends on ground surface condition. More is the vegetation cover higher will be the value of z_0. The friction velocity (u_*) is a measure of shear stress at the surface, and is used as the driving variable for windblown sediment transport in most wind erosion models. By setting anemometer at different heights (at least three) from surface, u_* and z_0 can be characterized for different surface conditions. During drought situation the aerodynamic roughness length will be very small since ground coverage in the form of vegetation becomes very low and therefore friction velocity will be very high near to surface to cause erosion. Considering the characteristic wind velocity profile at the ICAR-CAZRI, Jodhpur farmland with normal crops in the year 2014-15 having normal rainfall, annual average aerodynamic roughness (z_0) and friction velocity (u_*) as per Eq (7) were estimated as 0.25 m and 0.23 m s^{-1}, respectively, indicating fairly susceptible area prone to the wind erosion.

Wind velocity at three heights above surface (1 m, 2 m and 2.5 m) recorded at a rangeland site in Chandan during three windy situations, calm, moderate and high wind regimes are presented in Fig. 5 to show the typical wind speed profile. With increase in height from surface, wind speed increased rapidly up to 0.5 m and then increased gradually. In case of low wind regimes (bottom and middle line), wind speed near ground surface was almost nil. In case wind speed at 1 m height exceeds about 2.5 m s^{-1}, then only near surface wind speed will be greater than zero. It actually indicates the threshold wind speed at 3 m height, commonly measured at meteorological observatory to initiate the wind erosion process. And this threshold increases many folds at more number of occasions during drought ultimately increasing the menace of sand storms.

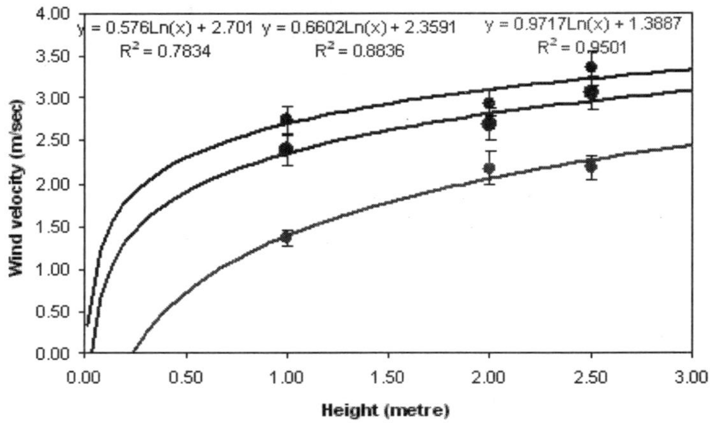

Fig. 5. Wind speed profile measured at Jaisalmer during three different windy periods

Wind Eroded Soil Loss as Affected by Factors

Factors of wind erosion process as affected by drought are discussed above, but how these factors contribute to wind eroded soil loss are presented below. In RWEQ model, all the above mentioned factors along with few more factors are related with soil loss through an empirical relationship as follows:

$$Q_{\max} = 109.8(WF \times EF \times SCF \times K' \times COG) \tag{8}$$

Where $Q_{\max}$ is the maximum aeolian mass transport rate capacity (kg/m-width), *WF* is the weather factor, *EF* is the soil erodible fraction factor, *SCF* is the soil crust factor, *K'* is the soil roughness factor, *COG* is the combined crop factor or ground vegetation factor ($SLR_f \times SLR_s \times SLR_c$) in which SLR_c indicates crop biomass factor. Soil loss is further estimated from $Q_{\max}$, detailed discussion of which may be found in RWEQ document (Fryrear et al 2000). Here, it is only shown that how drought affects *WF* and *COG* factors and further on aeolian mass transport rate. For example, during a drought situation at Jaisalmer if the wind factor reaches a value of 4 kg m^{-1} s^{-1} and there is no rainfall for continuous 15 days, the SW factor becomes 1, which results in *WF* factor as 4 kg m^{-1} s^{-1}. COG factors as mostly governed by flat residue (SLR_f) and standing residue (SLR_s) may be approximated as 0.8 and 0.15, respectively considering 5% flat residue on surface and standing residue as mentioned in Table 2. As compared to this drought situation, if W_f, *SW*, SLR_f and SLR_s are considered as 2 kg m^{-1} s^{-1}, 0.9, 0.33, and 0.10, respectively for a normal year, the $Q_{\max}$ will be 8 times higher in drought situation than normal year, if all other factors remain similar. It means during drought year, nearly eight times more sand will most likely be removed even if wind velocity is same during both, normal rainfall year as well as drought year.

EXTENT OF WIND EROSION HAZARD IN THE THAR DESERT

Kar (1993) estimated potential wind erosion index for Thar desert using mean wind speed and PE values of five months (March to July) from long term climatic data of the different stations in desert. Higher wind erosion index indicated more erosion hazard by wind the maximum being in Jaisalmer followed by Phalodi, Bikaner, Jodhpur, Pachpadra, Barmer and so on (Table 3). The above findings were substantiated by preponderance of erodible landscapes like dunes and hummocks in Jaisalmer, Phalodi and Bikaner. In addition, use of tractors has also accelerated the sand movement (Dhir et al 1992). The declining duration of fallowing or its complete absence has also reactivated sandy landscapes in both western and eastern parts (Kar, 1988).

Table 3. Categories of wind erosion index (March -July) for Thar Desert and meteorological stations in them (After Kar 1993).

Wind erosion index (%)	Category	Station
480 & above	Extremely high	Jaisalmer
120-479	Very high	Pahlodi
60-119	High	
30-59	Moderate	Bikaner, Jodhpur, Pachpadra, Barmer
15-29	Low	Ganganagar, Churu, Nagaur, Jaipur
1-14	Very Low	Hisar, Sikar, Sambhar, Ajmer, Erinpura

MITIGATION OF WIND EROSION HAZARDS

Such hazards of increasing dimensions during drought can only be minimised by providing vegetative cover on these habitats. Sand dunes stabilization, wind breaks and shelterbelts are proven techniques for mitigation and lessoning of these hazards.

Sand dune stabilization

Barren dunes and hummocks, which are often mobile, can be stabilized by vegetation cover. This decreases surface wind speed, prevents scouring action, ameliorates soil conditions and improve micro-climatic condition of the area. Species for sand dune stabilization should be able to survive in (i) extreme temperature conditions, (ii) a variety of salinity conditions, (iii) variable wind speed and direction, (iv) severe sand storm events, (v) very low soil moisture condition, and (vi) biotic stress situations. For achieving this, desirable adaptive traits in selected species are (i) rapid rooter (ii) succession facilitator; (iii) rapidly multiplies over time and space; (v) resilience to grazing or complete leaf removal or leaf fall; (vi) prolific seeder (vii) possesses staggered dormancy and longer seed viability (viii) commensals and as a nurse plant; (ix) economically important (x) resistant to pest and diseases. In fact at lower depths of unstabilized sand dunes considerable moisture (2-5%) even in the peak evaporation months of May-June ensures success of plantation by way of active root growth (Gupta, 1979).

Area for sand dune stabilization is fenced either by mechanical or vegetative fences. Then micro windbreaks in checker board pattern or parallel rows across wind (Harsh et al., 1995) are erected by way of brush wood from *Ziziphus nummularia*, twigs of *Prosopis cineraria*, *P. juliflora*, *Crotalaria burhia, Aerva persica, Leptadenia pyrotechnica* and *Calotropis procera*. Of late, live plants of *Cassia angustifolia* have emerged successful as micro windbreaks. This saves young saplings from abrasive action of sand creep. Suitable species like *Acacia trotilis, Prosopis cineraria, Prosopis juliflora, Calligonum polygonoides, A. nubica, Cordia rothii* and *P. cineraria* are then planted. Intervening spaces are planted/seeded with grasses like *Cenchrus ciliaris, Panicum turgidum, Lasiurus sindicus, Citrullus colocynthis* and *Saccharum bengalensis*.

Attempts have also been made to stabilize sand dunes by direct seeding through air craft. A mixture of seeds of trees, shrubs and creepers along with grass seeds in soil pellets were aerially broadcast on sand dunes in Bikaner during post-monsoon season of 1982-83. The germination and growth of *Acacia tortilis* and *Lasiurus sindicus* was maximum (Kumar and Sankarnarayana, 1988). The risks involved in this exercise were (i) drifting of seeds by wind during the broadcast (ii) falling of seeds and its trapping in the existing bushes and being lost (iii) rolling of seeds on the slope resulting in uneven distribution of saplings (iv) burial of seeds under sand thus preventing its germination (v) loss of seeds by zoochory and (vi) grazing of emerging saplings by both wildlife and livestock. Despite so many risks, aerial seeding can be extremely useful in areas receiving low rains and are inaccessible for ground based human assisted rehabilitation of sand dunes. The pellets of soil (clay + FYM + sand) in the above exercise were replaced with a polymer called 'Jalashakti' in a separate experiment. Seeds were treated with 'Jalashakti' @ 2 kg/100 kg seeds of different species. The survival in Jodhpur conditions after three years was 25% in *Colophospermum mopane*, 20% in *Acacia tortilis,* 15% in *A. bivenosa,* and 5-10% in all other species. The untreated seeds showed 1-10% survival. Thus, polymer treated seeds in direct broadcast for dune stabilization do have a potential (Harsh and Tewari, 1993). Such stabilization of dunes not only prevents sand drift, it can also be turned into an economic activity by way of providing 15-20 t/ha of wood after five years of plantation. Benefits in terms of fuel wood yield and pods in different plantation densities at different locations are assured, while a B:C ratio of 1.83 to 3.58 has been estimated (Kalla, 1977).

Windbreaks and shelterbelts

Three, five or seven rows of woody perennials mainly trees and shrubs plated in a specified geometry in `belts' across the wind are called shelterbelts. They reduce winds speed and stop erosion benefitting croplands in its lee zone

(Ganguli and Kaul, 1969). In order to be aerodynamic in shape, selected species should have optimum height, porosity of canopy and be pyramidal in shape. Central row of tall trees followed by side rows of small trees and/or tall shrubs with adequate porosity ensures smooth overflow of wind without much loss of energy. The effective area is 40 to 50 times the height in the lee side of shelterbelt where wind erosion hazard is minimised. Further details on their design can be seen in Mohammad et al (1996) and Cornelis and Gabriel (2005).

In Indian arid zone, however, five row shelterbelt in pyramidal shape with tall trees in central row (e.g. *Acacia, tortilis, Tamarix articulata, Azadirachta indica)* and smaller trees on two side rows of (eg. *Acacia senegal, Prosopis juliflora, Parkinsonia aculeata* etc.) and then again two rows of shrubs on either side [e.g. *Aerva javanica* (Syn. *A. tomentosa*), and *Calligonum polygonoides*] has been found suitable. The suitable species and their design for a variety of purposes in arid zone are given in Table 4. Farmers also use rock piles as mechanical wind barrier. Efforts have also been made to develop non-vegetative structures as wind barriers. Date-fronds mat fence successfully controlled wind erosion in Arabian desert (Murai et al1990 and Al-Afifi et al1990).

Table 4. Design and species for shelterbelts plantation

Purpose	Design	Suitable species
Road side	3 to 5 staggered rows	*Acacia tortilis, Tecomella undulata, Azadirachta indica, Dalbergia sissoo, Tamarix articulata, Parkinsonia aculeata, Acacia nubica.*
Railway side	3 to 6 rows	*Salvadora oleoides, Acacia jacquemontii, Parkinsonia aculeata, Acacia nubica*
Canal Side	4-6 rows	*Acacia nilotica, Eucalyptus* spp., *Azadirachta indica, A. tortilis, D. sissoo, P. cineraria*
Farm Boundary (Rainfed)	1/2/3 rows	*P. aculeata, Euphorbia triucalli, Euphorbia caducifolia, Ziziphus nummularia, Leptadenia pyrotechnica, Clerodendrum phlomoides, Calligonum polygonoides*
Farm Boundary (Irrigated)	1/2 rows	*A. lebbeck, Acacia nilotica, Eucalyptus* spp., *D. sissoo, Sachharum munja, Panicum antidotale, Ziziphus rotundifolia*

Wind screens of technical textiles also reduce the wind speed (Dierickx et al., 2001). In Tibet, gravel sand barrier successfully controlled wind erosion (Zhang et al., 2007). Even porous stainless steel fence effectively sheltered the saltating sand particles (Zhang et al., 2010) in a wind tunnel study. Attempts are underway to develop dual mechanical barrier at CAZRI, which will generate wind energy and simultaneously reduce the wind speed in leeward side of the barrier (Santra, 2012).

WAY FORWARD

Evidently, wind erosion process becomes severe when drought situation prevails. This is so because occurrence of drought creates the favourable condition for wind erosion process e.g. extremely dried loose sandy soil with minimum vegetation cover, high wind speed etc. The weather factor and surface cover factor as described in RWEQ are also linked with drought situation and can be directly quantified as per prescribed equations. Wind velocity profile can also be measured to assess the surface conditions that make it prone to erosion. Management of these surface conditions during normal rainfall year provides resilience to wind action during droughts as described above and this is the only way to mitigate it. More and more research is needed to find out novel plants, newer planting techniques and management of these areas with public participation for which policy actions are essential at all levels for ensuring success.

REFERENCES

Al-Afifi MA, Haffar I, Murai H , Itani S and Yokota H 1990 Use of date-fronds mat fence as a barrier for wind erosion control. 2. Effect of barrier density on microclimate and vegetation. *Agriculture, Ecosystems and Environment* **33** 47-55

Cornelis WM and Gabriel D. 2005 Optimal windbreak design for wind-erosion control. *Journal of Arid Environments* **61** 315-332

Dhir RP, Kolarkar AS and Singh HP 1992 Soil ressources of the Thar. In. Perspectives on the Thar and the Karakum (Eds. A. Kar, R.K. Abhichandani, K. Anantharam and D.C. Joshi), pp. 60-85. Department of Science and Technology, Govt. of India, New Delhi

Dierickx W, Gabriels D and Cornelis W 2001 A wind tunnel study on wind speed reduction of technical textiles used as windscreen. *Geotextiles and Geomembrane* **19** 59-73

Fryrear DW, Chen WN, Lester C 2001 Revised wind erosion equation. *Annals of Arid Zone* **40(3)** 265-279

Fryrear DW, Bilbro JD, Saleh A, Schomberg HM, Stout JE and Zobeck TM 2000 RWEQ: improved wind erosion technology. *Journal of Soil and Water Conservation* **55(2)** 183–189

Fryrear DW, Saleh A and Bilbro JD 1998 A single event wind erosion model. *Transactions of the American Society of Agricultural Engineers* **41(5)** 1369–1374

Ganguli JK and Kaul RN 1969 Wind erosion control. ICAR, New Delhi. *Techn. Bull. (Agric.),* **20** 1-53

Gupta JP 1979 Some observations on the periodic variations of moisture in stabilised and unstabilised sand dunes of the Indian Desert. *Journal of Hydrology* **41(1)** 153-156

Gupta JP, Aggarwal RK and Raikhy NP 1981 Soil erosion by wind from bare sandy plains in Western Rajasthan, India. *Journal of Arid Environments* **4** 15-20

Harsh L N and Tewari JC 1993 Sand dune stabilization, shelterbelts and silvi-pastoral plantation in dry zones. In: Desertification and Its Control in the Thar, Sahara and

Sahel Regions (Eds. A.K. Sen and A. Kar), Scientific Publishers (India), Jodhpur, India. pp. 269-279

Harsh LN, Tiwari JC and Suresh Kumar 1995 Sand dune fixation and afforestation in arid region pp: 171-178 In AK Sen and A Kar (eds.) *Land Degradation and Desertification in Asia and the Pacific region* Scientific Publishers Jodhpur

Harsh LN and Tewari JC 1997 Stabilization of sand dunes and their management in Indian arid zone. In: Desertification Control in the Arid Ecosystem of India for Sustainable Development (Eds. Surendra Singh and Amal Kar), Agrobios, India, pp. 317-325

Kar A 1993 Aeolian processes and bedforms in the Thar Desert. *Journal of Arid Environments* **25** 83–96

Kar A, Felix C, Rajaguru SN and Singhvi AK 1998 Lateholocene growth and mobility of a transverse dune in the Thar desert. *Journal of arid environnements* **38** 175-185

Kalla JC 1977 Statistical evaluation of fuel yield and morphological variates for some promising energy plantation tree species in western Rajasthan. *Annals of Arid Zone* **16(1)** 117-126

Kumar S and Shankarnarayan KA 1988 Aerial seeding on sand dunes: Seedling survival and growth. *Journal of Tropical Forestry* **4(2)** 124-134

Mertia RS, Prasad R, Gajja BL, Samra JS and Narain P. 2006 Impact of shelterbelts in arid region of western Rajasthan. Central Arid Zone Research Institute, Jodhpur, India, pp. 76

Mertia RS, Santra P, Kandpal BK and Prasad R 2010 Mass-Height Profile and Total Mass Transport of Wind Eroded Aeolian Sediments from Rangelands of Indian Thar Desert. *Aeolian Research* **2** 135-142

Mohammed AE, Stigter CJ and Adam HS 1996 On shelterbelt design for combating sand invasion. *Agriculture, Ecosystem and Environment* **57** 81-90

Murai H, Al-Afifi MA, Haffar I and Yoshizaki S 1990 Use of date-fronds mat fence as a barrier for wind erosion control. 1. Effect of barrier density on sand movement stabilization. *Agriculture, Ecosystems and Environment* **32** 273-282

Ramakrishna YS, Rao AS, Singh RS Kar A and Singh S 1990 Moisture, thermal and wind measurements over selected stable and unstable sand dunes in the Indian desert. *Journal of Arid Environments* **19** 25-38

Ramana Rao BV, Sastri ASRAS and Ramakrishna YS 1981 An integrated scheme of drought classification as applicable to Indian arid region. *Idojaras* **85** 317-322

Rao AS and Vijendra K Boken 2005 Agricultural Drought Monitoring and Management in India (Chapter 23) In: *Monitoring and Prediction of Drought: A Global Stud*y (Eds: Vijendra K. Boken, A.P. Cracknell and R.L. Heathcote) Oxford University Press, New York, pp 297-312

Rao AS, Gupta JP and Faroda AS 1997b Changing climatic scenarios and strategies for drought management in the Indian arid region. *Drought Network News*, Nebraska **9(1)** 8-11

Rao AS 1997 Impact of Droughts on Indian Arid Ecosystem. In: *Desertification Control in the Arid Ecosystem of India for Sustainable Development* (Eds: Surendra Singh and Amal Kar), Agro Botanical Publishers (India), Bikaner, pp 120-130

Saleh A 1993 Soil roughness measurements-chain method. *Journal of Soil and Water Conservation* **48 (6)** 527-529

Saleh A 1997 Reply to "Comment on chain method for measuring soil roughness". *Soil Sci. Soc. of Amer. J.* **61 (5)** 1533-1534

Santra P 2012 Development of dual purpose mechanical barrier to control wind erosion with simultaneous utilization of renewable energy. DST funded project under Fast Track Scheme for Young Scientist (Project code SR/FTP/ES-60/2011), DST, Govt of India

Santra P, Goyal RK, Tewari JC, Singh JP and Roy MM 2013 Assessment of potential soil loss rate by wind and water erosion in Jodhpur region of western Rajasthan, India. In: GlobalSoilMap, Basis of the global spatial information system (Eds: D. Arrouays, N. McKenzie, J. Hempel, A. Richer D. Forges, A. McBratney), CRC Press: Taylors and Francis Group, London, pp. 139-143

Skidmore EL, Hagen LJ Armbrust DV Durar AA, Fryrear DW, Potter KN, Wagner LE and Zobeck TM 1994 Methods for investigating basic processes and conditions affecting wind erosion. *In : Soil Erosion Research Methods* (Ed. R. Lal). *Soil and Water Conservation Society, Ankeny*. IA, pp. 295-330

Wagner LW and Yu Y 1991 Digitization of profile meter photographs. *Transactions, ASAE* **34 (2)** 412-416

Woodruff NP and Siddoway FH. 1965 A wind erosion equation. *Soil Science Society of America Proceedings* **29** 502-608

Zhang CL, Zou XY, Cheng H, Yang S, Pan XH, Liu YZ and Dong GR 2007 Engineering measures to control windblown sand in Shiquanhe town, Tibet. *Journal of Wind Engineering and Industrial Aerodynamics* **95** 53-70

Zhang N, Kang JH and Lee SJ 2010 Wind tunnel observation on the effect of a porous wind fence on shelter of saltating sand particles. *Geomorphology* **120** 224-232.

23

Socio-Economic Impacts of Drought in India

Khem Chand[1] and Nagaratna Biradar[2]

INTRODUCTION

Aberrant weather is a part of agriculture in India affecting its production. Drought and floods are the result of this aberrant weather making agriculture one of the weather sensitive enterprises. Floods, cyclones and storms have their adverse impact in a decreasing order of magnitude in view of limited area of occurrence. But, drought is extensive and prolonged. As a result drought is considered as one of the biggest menace to agriculture among all weather related crisis.

Past records indicate that in almost every year one part or other of the country has been subjected to drought, flood or cyclone. Rain-dependent areas can be broadly split into two: 'dry lands', which receive less than 750 mm of rain a year; and rainfed areas, which receive more than 750 mm. Comprising arid and semi-arid ecosystems, dry lands stretch from Gujarat in the west till Eastern Madhya Pradesh; and from Rajasthan till the southern tip of India. The problem of drought, especially agricultural drought is more specialized and complicated (Palmer 1965) with many impacts/consequences on farming community.

MEANING OF THE TERM 'IMPACT'

The dictionary meaning of the term 'impact' is the action of one object coming forcibly into contact with another. It also means a marked effect or influence. So, in simple terms it is a measure of the tangible and intangible effects (consequences) of one thing's or entity's action or influence upon another. The thing or entity is 'drought' and another thing here refers to 'rural area' as we are deliberating on impacts of drought on rural area.

---

1. Principal Scientist and Head, 2. Scientist
ICAR-Indian Grassland and Fodder Research Institute, Jhansi (UP)
Email: kcmamnani@gmail.com

MEANING OF SOCIO-ECONOMIC IMPACT

Socio-economic impact means influence of a thing/process on social factors, economic factors and on interaction process of both social and economic factors together. Socio-economic factors are the social and economic experiences and realities that help mold one's personality, attitudes, and lifestyle. The factors can also define regions and neighborhoods. In general socio-economic impact analyses how societies progress, stagnate or regress because of influence of a particular thing.

'Thing' here refers to drought and hence we need to understand the meaning of 'drought'.

MEANING OF DROUGHT

Drought results from long continued dry weather and/or insufficiency of rain, which causes exhaustion of soil moisture, depletion of underground water supply and reduction of stream flow. Drought is frequently defined according to disciplinary perspective. Bandyopadhyay (1988) lists four types of droughts, namely (1) meteorological drought, (ii) surface water drought, (iii) ground water drought and (iv) soil-water drought. He argues that the various forms of droughts get generated independently but inseparable and are linked to each other through the water cycle. The National Commission on Agriculture in India defines three types of droughts, namely, meteorological, agricultural and hydrological droughts. Meteorological drought is defined as a situation when there is significant decrease from normal precipitation over an area (i.e. more than 25%). Agricultural drought occurs when soil moisture and rainfall are inadequate during the growing season to support healthy crop growth to maturity and causes crop stress and wilting. Hydrological drought may be a result of long term meteorological droughts which result in the drying up of reservoirs, lakes, streams and rivers, and fall in groundwater level. Many others have also included economic or socio-economic factors. **Social drought** relates to the impact of drought on human activities, including indirect as well as direct impacts: and **Economic drought** – "a meteorological anomaly or extreme event of intensity, duration (or both), outside the normal range of events that enterprises and public regulatory bodies have normally taken into account in their economic decisions and that, therefore, results in unanticipated (usually negative), impacts on production and the economy in general". The concept of drought varies from place to place depending upon normal climatic conditions, available water resources, agricultural practices and the various socio-economic activities of a region. The various approaches taken by scientists and non-scientists to define drought demonstrate its complex and interdisciplinary nature. At the same time, although most definitions emphasize the physical aspects of drought, the social aspects are closely related.

Drought thus has many definitions and meanings for different stakeholders. For farmers, it simply means lack of adequate moisture for normal crop growth.

Impacts of drought

Majority of farmers in the country follow rainfed farming. In India, about 60% of total net sown area comes under rainfed lands. Rainfed crops account for 48 percent area under food crops and 68 percent under non-food crops. The climate in India's rainfed regions is characterized by complex climatic deficiencies, manifested as water scarcity for rainfed crop production. The climate is largely semi-arid and dry sub-humid with a short (occasionally intense) wet season followed by long dry season. The farmers of this region are the ones who are more prone to drought related effects. Drought at any stage of crop growth has profound effect on them. The socio-economic effects are discussed in this chapter keeping this large chunk of rainfed farmers in focus. The impact of drought varies with the duration of droughts. The longer the period of drought and the larger the number of consecutive droughts, the greater will be its impact on agriculture, ecology and economy. Understanding impacts on these is however not important in the context of this chapter as the focus is on socio-economic aspects.

The impact of a drought on the overall economy of the country is evident at macro (State and national level) and micro–levels (village and household levels). The impacts could be direct or indirect and vary in nature and intensity. The extent and intensity of drought impacts is determined by prevailing economic conditions, the structure of the agricultural sector, management of water resources, cereal reserves, internal and external conflicts, etc (Benson and Clay 1998).

DIMENSIONS OF SOCIO ECONOMIC IMPACT OF DROUGHT

At macro level where economy at large is affected: In the industrial sector, agro-based industries are the ones directly affected, as the lower domestic production of agro-processing inputs reduces non-agricultural production. Domestic availability of water is also restricted. This latter aspect has implications for health and household activities, including the time required for collecting water. As water becomes scarce, competition among and within sectors may increase. Droughts have potentially important implications for government policies, first and foremost, via their impact on the budgetary balance. A drought is likely to reduce tax revenue via a decline in income, employment and exports. On the expenditure side, increased expenditure on relief, social welfare, health and water supplies, consumption-related subsidies on food distribution, and the logistical costs of drought-related imports may confront government. Law and order services could also be put under greater pressure by a rise in crime, in turn associated with temporary unemployment, migration and increased destitution. In addition, there are likely to be pressures for the increased provisions of subsidies and credit to affected productive sectors, including public utilities. Government policies can exacerbate the impacts of drought.

At micro level where in farm household is the unit: Micro level impact is largely on the entitlement to produce and procure food. The impact varies depending upon the social structure (class, caste) and village and household resource endowments. Drought entails a loss of assets in the form of crop, livestock and productive capital damage as a direct consequence of water shortage or related power cuts. Micro level impacts are discussed in detail in subsequent part of this chapter.

IMPACT OF DROUGHT AT MICRO (VILLAGE AND HOUSEHOLD) LEVELS

In general, three livelihood sources 'crop, livestock and labour' dominate in villages. Each household may derive its livelihood either from one of it or combination of these sources. Extent of contribution of each source on the livelihood of villagers varies for different categories of farmers. While resource rich farmers may derive most of their income from crop and livestock, the poor farmers may derive large part of their income from livestock and labour. There may be some other sources of livelihood like petty shops, artisans etc, but these three sources predominate in rural areas. The socio-economic impacts of drought thus are discussed keeping these three sources as major factors.

Impact of drought on livelihood source- Crop husbandry

Drought impacts crop husbandry, one of the main livelihood source bringing in more specific and inter-linked impacts as given in Figure 1.

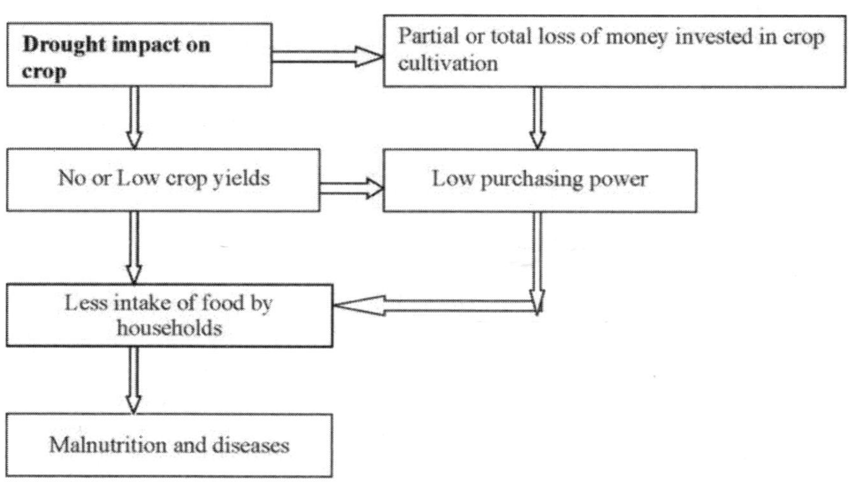

Fig. 1. Drought impact on crop husbandry

These specific impacts are discussed here.

In past, the rainfed farming system was mainly dependent upon the locally available inputs (seeds, manures, animal draft) and used to grow a number of crops, which were able to withstand drought-like situation. However, in recent times, the cropping systems have changed and currently the farmers in these rainfed areas have limited options. Many of the farmers in these regions started cultivating high value crops which require intensive use of costly inputs (chemical fertilizers/ pesticides, hybrid seeds, life saving irrigation, farm energy etc.) and find it difficult to manage the resources on their own.

So, today's agriculture is input intensive. Hybrids are grown as they yield more. Traditional method of saving seeds is no more a practice in the villages. Land preparation is in most cases by hired tractor. Chemical fertilizers are must when seeds are sown. So, money is now the essential input for the farmer to begin the crop season. The loss due to crop yield thus is two folds- loss of initial investment (partial or total) and loss of crop harvest. Majority farmers for the initial investment depend on credit sources and non repayment due to crop loss leads to piling up of interest. The growing loan amount becomes huge financial burden for the farmers pushing them into vicious circle of poverty.

Besides, farmers are faced with an inadequate harvest to feed their families and to fulfill their other commitments. Low intake of food causes malnutrition, hunger and starvation that poses threat to health and spread of diseases like diarrhoea, dysentery, cholera and opthalmia. Ravindranath et al (2005) in their study on effect of drought on nutritional status of rural community in Karnataka found that the intakes of foodstuffs at the household level, such as green leafy vegetables, roots and tubers, other vegetables, fruits, milk and milk products, fats and sugar was lower as compared to recommended dietary allowances (RDA) as well as the levels reported by NNMB in its survey during non-drought period (2000-2001). The daily average intake of all the nutrients except calcium and thiamin was less than the levels recommended. The adverse effect of drought was reflected by marked decrease in the intake of vitamin A and concomitant increase in the prevalence of Bitot's Spots among preschool children. Food insecurity, particularly with respect to income elastic foods was significantly higher among SC/ST communities and in the households engaged in labour activities. Similarly, higher proportion of households of SC/ST communities and those households engaged in agricultural labour, adopted coping strategies like consumption of low cost foods and reduction in the quantities of food consumed.

Udmale et al (2014) reported that about 60% of respondents faced high to very high food scarcity during the drought years as compared to the normal years in Maharashtra and 52% of respondents said that drought has threatened their house- hold food security to a greater extent. Approximately 58% of respondents agreed that they have less to very less food grain choices for their daily consumption.

The loss of crop in one season thus has implications on the life of farm families for many years.

Impact of drought on livelihood source - livestock husbandry

Specific impacts of drought on livestock husbandry are given in Figure 2.

Common property resources (CPRs) such as open grazing lands are decreased or not available for several villages today. There is evidence indicating a rapid decline in CPRs, both in size and productivity (NSSO, 1999). Farmers graze their livestock on roadside and on field bunds. The little native grass available there in good rainy season becomes unavailable during drought. The crop loss will be either partial or full depending upon the severity of the drought. Complete crop loss becomes catastrophic for the livestock as they are left with no feed (green or dry) almost throughout the year. The partial loss of crop however provides hope though little for farmers as the stunted crop can be used as fodder for these animals for a few months at least in a year. Whatever may be the situation, livestock are inadequately fed during drought leading to their poor health and extremely low production.

The situation of water availability to livestock also becomes bleak during drought. Lack of water harvesting structures and poorly maintained ponds (with silt deposition) pose challenge to farmers to meet water requirement of livestock. Yield of bore-wells (tube wells) reduce so much so that sometimes it may not be sufficient to meet human requirement leave aside the livestock. The little quantity of water even if available to livestock many times is contaminated as farmers are forced to use the same both for livestock drinking and bathing. It seriously affects livestock health making them prone to diseases.

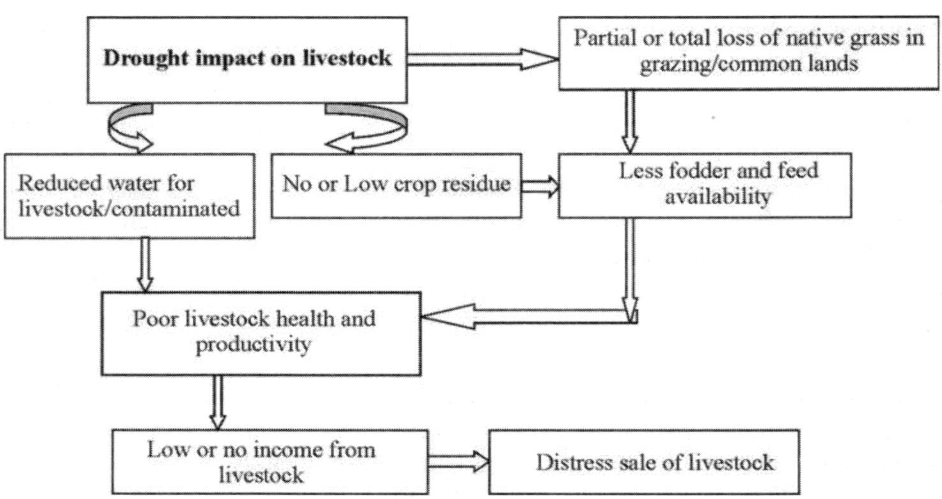

Fig. 2. Drought impact on livestock husbandry

Inadequate availability of fodder and water to livestock affects their health and obviously their production level too. The shooting price of fodder (almost double or triple) makes it unaffordable for many small and marginal farmers who own large chunk of livestock.

A study by Biradar and Sridhar (2009) revealed that in normal year 50.92 % of farmers of Karnataka purchased fodder and feeds. This was increased to 81.18% during drought year. The t-test calculated showed significant difference in the number of farmers purchasing fodder during normal and drought years. While majority of the farmers purchased concentrate feeds in normal year, during drought farmers purchased crop residues also. In normal year, none purchased sorghum stover, but in drought year nearly half (43.24%) of the respondents purchased it. The shortage of fodder was caused largely by outflow of rabi Jowar stocks from Karnataka to neighboring Maharashtra and Tamil Nadu. Purchase of paddy straw by the number of farmers was more than doubled in drought year (37.71%) as compared to normal year (16.97%). On the contrary, number of farmers purchasing concentrate feed reduced to half in drought (23.61%) from 46.49% in normal year. Similar trend was observed for the purchase of rice bran as number of farmers purchasing it reduced from 28.78% in normal year to 17.71% in drought. None purchased hay of legume crops and other crop residue in normal year. But in drought, 8.85 per cent of them purchased the former and 9.22% the latter. This indicates that maintenance feed of the livestock was though addressed by the farmers but with an extra cost. In a drought situation, with the non-availability of adequate and quality fodder, providing cattle feed to cover a part of the daily requirement of animals is essential. The increase in the cost of feed almost by 30 percent, especially by Karnataka Milk Federation (DARE 2003) due to hike in the price of raw materials aggravated the situation further affecting livestock health and productivity.

The average quantity of fodder and feeds purchased by the farmers during normal year was 22.83 q and increased to 38.80 q in drought year, up by 69.95%, in spite of increase in the cost of fodder and feeds. Significant difference was hence found in the quantity of fodder and feeds purchased during the years under comparison.

The value of fodder per ton including cost towards labour and transport worked out was Rs. 2199 in normal year that was almost doubled (Rs. 4166) during drought years. According to DARE report, the fodder prices in Karnataka went up to 75 per cent higher than the normal during November 2002, February 2003 and peaked again in May-June 2003. In many cases the cost of transportation exceeded the actual cost of fodder. For instance in Chamarajnagar district, they purchased paddy straw from the Kabini command area, situated next to the district, where it was available plenty at cheaper rate. But increased expenditure on transportation of fodder enhanced its overall cost. The farmers traveled average of 4.25 km radius to purchase the fodder in normal year. This distance traveled in drought years increased 10 times as they had to travel an average distance of 44.07 km.

Not able to maintain and to earn from livestock means that the only option is to go for distress selling at an incidental price. In Mahabubnagar, in the three villages of the study area, 78.2% of cattle, 93.5% of buffaloes, 88.8% of cows, 84.3% of sheep, 36.3% of goats and 100% of calf population owned by different categories of households were sold away in distress. Different prices of livestock prevailed during the drought in the seven villages of Anantapur and Mahabubnagar. The prices of livestock during drought distress are reported to be on an average about fifty percent than in normal times. It is not uncommon for many small farmers to sell their draught animals in distress and again get into debt to buy similar stock at higher price once the monsoon starts. Purchase of cattle is an important cause for many small farmers indebtedness.

Distress selling of livestock during drought was resorted to by 17.34 per cent farmers. An average of 2.55 ACU were sold to the middlemen of slaughter houses at throw away price. However majority of the farmers resorted to distress selling were marginal and small farmers. Livestock for these farmers acts as an important livelihood source and devoid of fodder in common property resources, on which they heavily depend, affected them (Biradar and Sridhar 2009).

Rathore (2004) reported that in Rajasthan sheep and camels, followed by cows, were the most vulnerable animals. They perished in large numbers ranging between 10 to 32% of the population. Primary surveys have shown that losses of cows were much higher, as large numbers i.e., 25 to 50% of cows were abandoned due to starvation or put in charity centers, because of acute shortage of fodder and finances. The remaining cows became dry and weak. There was nobody to buy cows even for slaughter, as there is a ban on cow slaughter in Rajasthan. The severity of drought can be judged by the fact that even high price animals like buffaloes suffered and a few perished. Goats and sheep are also affected by drought, but relatively mildly compared to large animals. They are mainly grazing animals and depend on common property resources i.e., grazing lands, wastelands, forests etc. Moving to better-endowed areas saves them. Migration of small animals is a traditional practice in Rajasthan. In case of widespread drought, they are taken out of the State, and in the worst situation sold for slaughter, as there is a round the year market for them. All caste, class and economic categories of household keep small animals for milk and ready cash. However, their health is affected by loss of vegetation of their choice. When rain comes, there is a change in type of vegetation for grazing and few animals cannot adjust and die (10% of the total number, as reported by people). The drought year 2002-03 was bad for sheep as there was widespread drought in Rajasthan and also in adjoining States, therefore even the migration strategy did not work and more than 32% of sheep perished.

So, the impact of drought on livestock is manifested in terms of mortality, loss in productivity, health of animal and loss in fertility. Distress sale or mortality of livestock in particular brings huge loss to small and marginal farmers as buying livestock again means a saving of money for many years or looking for credit options. Besides, livestock provides partial employment to family labors and lack of livestock also means lack of work opportunity to them.

However, it should be noted that in areas where crops and animals occupy the same ecological zone, farming is likely to be a more risky business, because livestock compensate for localized rainfall shortages by movement to better favored areas.

Impact of drought on livelihood source -Labour

Drought impacts farm labors in many ways. The impacts and their linkages are depicted in Figure 3.

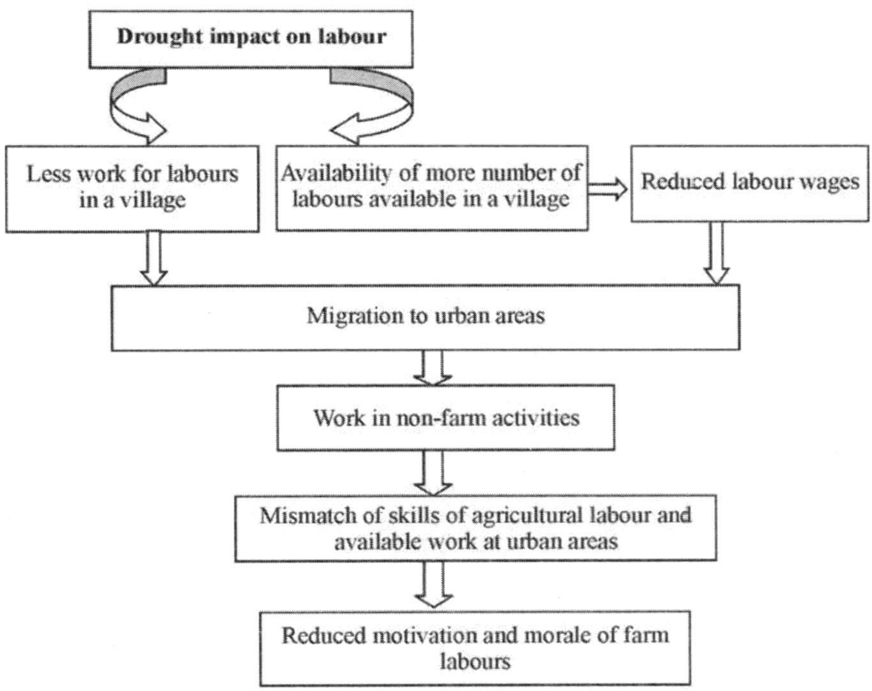

Fig. 3. Drought impact on farm labours

Two important sources of income for farmers are crop and livestock. Results of the average annual income, defined as net returns to family owned resources which include family labour and livestock, capital and land, of farmers during normal and drought years indicated overall reduction of annual income by fifty per cent during drought year in Karnataka state (Biradar and Sridhar 2009).

Percent reduction was more in case of income from crops (61.42%), followed by livestock (30%) and least by others-labour (20%). Bidinger et al (1991) reported similar findings that due to drought there was sharp fall in the share of crop income. Sources of crop revenues literally dried up as the share of crop income fell from 58% of net household income before the drought to 7% in the second drought year. However the effect of income reduction due to livestock was more pronounced for small and marginal farmers. While annual income of 75.64% farmers in normal year was more than Rs.11500 (ceiling fixed by Department of revenue, GoK for families below poverty line), in drought years only 31.36% of farmers reported that their annual income crossed above Rs.11500.

Pandey and Bhandari (2006) reported that in Chhattisgarh, Jharkhand and Orissa, almost 13 million people who sit perilously just above the poverty line fall back below it due to drought-induced income loss. The total farm household income can drop by 40 to 80 percent in drought years relative to normal years.

Two important income sources- crops and livestock are jeopardized in drought years as discussed above. The option thus left for them is to look for other jobs to earn living. The skills they have are of raising crop and livestock. In drought years, number of available labours is more than the number required for the jobs in a village. After having a look at the employment situation in the four villages, we find that about 75% of the males and females on an average are able to get employment opportunities in the village itself or in the surrounding areas for a period of not more than 90 days on an average in a year during the drought period. In such situations farmers are forced to move to far flung places in search of jobs to which they are not familiar with. Roy and Hirway (2007) reported that the drought prone areas of Gujarat have traditionally been an area of push factor for out migration in search of employment and livelihood. Seasonal impact on livelihood pushes large chunk of population out of this area to augment their sources of livelihood. More than 17 per cent of the total household surveyed reported at least one family member out-migrating in search of a job to sustain life of their family members. This leads to mismatch of skills for the jobs available elsewhere and force them to do unskilled labour work in harsh conditions for minimal wage. More labour availability means lesser wages in labour market. This forceful situation of taking low wages affects ones motivation to work and morale. This has the potential danger of leading farmers to addiction to alcohol and the consequences of alcohol addiction on one's own life and life of his family are calamitous.

Changes in the distribution of wealth

The experience of different farm households in times of drought will be determined by their ownership of assets, access to incomes from other sources and the extent to which these assets and incomes are less affected by drought than are harvests. The most vulnerable amongst those hit by drought will be those with few assets of value to sell, those who most need to purchase grain due to an absence of their own household reserves, and those who cannot gain access to food through other means, like borrowing, coercion or theft. Wood (1976) shows in his study of farmers in northern Ethiopia over the drought years of the early 1970s that the speed with which farmers became seriously short of food varied very considerably, due to their differing access to income sources. The richest farmers were even in a position to benefit by acquiring land and other assets from distress sales by their poorer neighbours. Sen (1981) formalises the analysis of differential impacts on incomes and wealth in times of crisis in his essay on "Poverty and Famines". He introduces the idea of "food entitlement an individual's overall access to food depending not only on his direct crop production but also on his access through the market - by sales of labour and other commodities - and through non-market channels such as social networks and re-distributive systems. Relative price movements change the entitlements of certain groups, making it more difficult for them to satisfy their food needs. Many farmers are in a similar situation, need to sell or mortgage land, jewellery and personal property to increase their power to purchase foodstuffs. A typical rural household in India spends about 15% of its annual income on celebrating festivals. But, drought mostly affected the income of low income farmers forcing them to reduce their expenses on festivals, which has a negative impact on social life and mental health. Nearly 54% of respondents reported high to a very high reduction in their expenses on festival celebrations (Udmale et al 2014). There was a significant difference observed in spending on festivals based on income groups with poorer farmers affected the most. Also, farmers from low income groups reported that their expenses on wedding ceremonies were reduced or ceremonies were postponed due to drought. A significant difference was observed in perception of conflicts for water in the society due to drought based on farmers' education level. Less educated farmers said that drought driven water scarcity caused conflicts in the society. Farmers from frequent and severe drought affected areas believed that drought was one of the major causes of suicidal tendencies of farmers due to less income and indebtedness with growing interest. Due to reduced income and social status farmers become mentally depressed, which ultimately ends in suicide. Rathore (2004) conducted a vulnerability assessment in Rajasthan and documented vulnerable population in four climatic regions of the State by caste, income and farm size categories. It was shown that both low-income and middle-income households are vulnerable to droughts. In the context of farm sizes, it is marginal and small farmers that are normally most vulnerable, whereas in desert region, even the large land holders are vulnerable due to the inferior quality of land with no irrigation facilities and direct dependency on erratic rainfall. Landless households are better off as they are less dependent on agriculture and livestock for their livelihood and more mobile.

These impacts deliberated above resulted into social, health and psychological impacts on farmer's livelihoods. It involved impacts such as inequities in the distribution of water or conflicts between water users, reduced school attendance, population migration, poor health and hopelessness or a sense of loss due to drought. Study of Udmale et al (2014) revealed that about 56%, 60%, 70%, 65% and 83% respondents rated these impacts to high to a very high extent, respectively.

Adaptation and mitigation measures

Loss due to drought is certainly scary as it shakes the very livelihood of farmers. But like any other human being, farmers are also adaptable to the loss. If not, the loss caused will intensify farmers' reaction for fear and panic. Positively it can also cause one to look at future and think of some adaptive measures to sustain the loss and survive. Major change is triggered in them when farmers face a significant discrepancy between what they expected and what actually happened due to drought. Farmers of arid and semi-arid region having experienced frequent droughts have learnt some measures to keep walking forward, even if it sometimes take an uncomfortable amount of effort to do so. They explore, learn, grow, accept difficulties, adapt and still carry on with farming. Three types of energy are required to make adaptations to change- mental (to figure out what is happening and how to respond) emotional (to deal with various feelings like loss, anxiety, threat, etc.) and physical (to accommodate the bodily implications of stress, excitement, etc.). Government through its various programs enables farmers to adapt to the changes due to drought. So, government programs act as enablers for the farmers to face the difficulties encountered. Adaptation to drought is a two-step process, which initially requires the perception that drought is occurring and then responding to its various impacts through adaptation and mitigation activities. Results of a recent study conducted by Udmale et al (2014) at Maharashtra to know the adaptation strategies followed by farmers and government to drought are taken as a case and given in these sub-sections.

Adaptations by farmers

Based on the perception of drought impact severity, farmers used various drought preparedness and adaptation measures to mitigate the drought impacts. Drought mainly affects the crop and livestock production, therefore, about 78.8% of farmers preferred not to sell their crop produce, and instead they stored it to deal with anticipated droughts. About 47.9% of farmers stored crop residues to fulfill the fodder demand during the anticipated drought, and 51% of farmers reduced their expenses and saved money. Farmers seek various options such as migration for employment, selling of livestock, non-agricultural income sources to lessen the drought impacts. Only few farmers having irrigation facilities sow crops on time in spite of irregular monsoon and even less choose crops requiring less water to deal with drought. However, it is observed that farmers from less irrigated areas tend to be well prepared to deal with anticipated drought by storing harvested grain and saving money as compared to medium- and high-irrigated areas. Farmers with small land holding tend to sell their livestock and seek alternative sources of income to cope with drought.

Based on the perceived severity of drought impacts, farmers adopted a variety of autonomous adaptation strategies to mitigate drought impacts on agriculture. Major agricultural adaptations identified by study are changing the crop calendar, using low water consuming crops, no sowing, using improved irrigation practices, water harvesting and reducing wastage of water during drought. It was found that about 61% and 53% farmers gave high preference to changing their crop calendar or adjusting cropping dates and using drought tolerant less water consuming crops respectively to mitigate the drought impacts. About 53% farmers highly preferred not to sow crops if the soil moisture is insufficient for a successful crop. During drought years, about 56% farmers reduced wastage of water to high to very high extent. These adaptation practices were widely used as these practices do not need extra financial cost and are easier to implement.

In the case of rainwater harvesting through various structures and use of modern irrigation practices such as sprinkler and drip irrigation, farmers' preference for this response was low. Only 40% of farmers rated as high or very high use of water harvesting and in-situ water conservation practices, whose use is seriously constrained by costs associated with them. The use of modern micro-irrigation practices such as sprinkler- and drip- irrigation was also not popular, due to high initial investment, high cost involved in renewing systems and lack of irrigation water source. Use of this micro-irrigation technology by farmers has been reported very low as compared with the potential this technology offers and the main reasons for this are constraints associated with obtaining, understanding or maintaining the technology, household income, farm size, and power supply constraints.

Land holding size, household income and education are among factors that significantly influence farmers' behavior with respect to adaptation strategies. Also, irrigation area, frequency and intensity of drought are other factors possibly influencing adaptation strategies of farmers. It was found that farmers with high land holding size highly preferred to change their crop calendar and used low water consuming crops against farmers with marginal and small land holdings. A significant difference was observed in adoption of adaptation practices such as use of sprinkler and drip irrigation practices from groups such as irrigation strata, landholding size, household income and drought intensity. This clearly indicates that the use of these adaptation strategies is highly affected by availability of water for irrigation, farm size and income. Also, farmers facing frequent drought of severe intensity were more likely to take steps to reduce their vulnerability to drought before it occurred by adopting various water management practices such as micro-irrigation practices. Farmers with high income and high land holding size concerned about management of available water without wasting it. It was observed that, in spite of good perception of severity of drought impacts and familiarity with various adaptation options, the preference given for their adoption in agriculture was not good enough.

MITIGATION MEASURES BY GOVERNMENT

Besides household level adaptation measures, administrative strategies play a very crucial role in adapting to drought. As a response to serious drought events in the Maharashtra State, the government has undertaken various relief measures, which included provision of employment, supply of drinking water and distribution of fodder in cattle camps. In addition to this, the government has also provided agricultural loans with low interest rates, crop insurance schemes, and waived electricity bills depending on intensity of drought.

Major drought relief measures undertaken by Government of Maharashtra during drought 2012–2013 are given below-

Employment scheme: The National Rural Employment Guarantee Act (NREGA) was introduced in 2005 aiming to provide 100-days employment to adult members of any rural household, who are willing to do unskilled manual work at the wage rate fixed by the government every year. In the DPA, the scheme is designed to serve mainly two purposes – first is creating employment opportunities for rural households affected by drought and second is building drought resilience. Under this scheme, drought proofing activities such as water conservation, increase in water harvesting potential of ponds and reservoirs in terms of manual excavation and construction of tanks, check dams, percolation tanks, underground dykes, ponds, rain water harvesting structures etc. are undertaken in DPA. Along with these activities, it also includes activities such as watershed development, tree plantation; labor intensive fencing, nursery raising, canal development, renovation of traditional water bodies and other related activities. This scheme is considered as one of the administrative level adaptation and mitigation measures. For the questions related to the employment scheme, mixed types of responses were received from the farmers. Only 46.2% of respondents knew about work started under NREGA from Gram Panchayats (GPs: local self-governments at the village or small town level in India), the rest of them did not receive any information or did not know anything about it. Approximately 75.7% of respondents having information about NREGA demanded for employment under the scheme, while only 41% of them got opportunity to work.

An average male and female labor wages per day were 2.82 and 2.19 USD respectively. About 56.3% of respondents rated the labor wage as insufficient, while 43.7% of them said the amount was just enough to fulfill their household needs. This opinion differs among households depending on number of family members, employment condition and living standard. Problems such as lack of desired and timely employment, low and untimely payment of wages, corruption, fake jobs and lack of proper NREGA implementation guideline were reported by the respondents. Beneficiary household's satisfaction rating of the employment scheme was very less.

Tanker water supply: Prolonged drought always results in increased dependency on groundwater resources. Excessive exploitation of groundwater resources results in depletion of groundwater level and drying of bore wells. These kinds of problems are observed very frequently in the DPA, which threatens the household drinking water supply. However, water is made available to villages and hamlets through private and government tankers depending on intensity of drought and villagers need. In this situation, agricultural and industrial water use was controlled. In 2012–2013, Maharashtra government deployed 1730 water tankers for 1356 villages and 3938 hamlets until February 2013. About 56.5% of respondent households were provided with tanker water supply under government drought relief measures. Level of satisfaction from the tanker water supply scheme was very less amongst the respondents, because water-supplying tankers did not have a fixed schedule, water supply was irregular (once in two days to once in two weeks), both quantity and quality of water was insufficient, and inequity in water distribution among households. To avoid conflicts among households, many village GPs dropped water into a public open dug well, from which households took water with the help of rope-bucket. But it resulted into excessive labor and even accidents while taking water out of the dug well with rope bucket. Community suggested dropping water from tankers into existing elevated public water supply tanks, and distributing through pipes for fixed duration to all households as a possible solution to the aforementioned problem.

Cattle camps: Water- and fodder- scarcities are the major issues for livestock management during drought. At the very early stages of drought, fodder distribution depots were assigned to private investors in the villages. They transported fodder from some fodder rich areas and distributed it amongst the needy farmers at low price. But as the drought prolonged, these cattle fodder depots converted into mass cattle camps and supplied free fodder and fodder supplements to the livestock. The state provided USD 1.5 and 0.7 per day to cover fodder, transportation, water supply and other related facilities costs for a big and small animal, respectively, accommodated in the cattle camp. According to Government of Maharashtra, about USD 46.6 million was spent on cattle camps until 15th August 2012 and USD 37.9 million was projected until June 2013. The projected expense for Ahmednagar and Solapur district was very high as compared to other districts (most of these district areas falling in the Upper Bhima catchment). Despite of ineffective implementation of the relief measure, respondents agreed that it was better than having nothing. Respondents' level of satisfaction to fodder depots and cattle camps as relief measures was medium on 5-point Likert type response. The major drawbacks of cattle camps highlighted by respondents' were allocation of less fodder to the cattle violating the government rules, no or less water availability for cattle, and no shadow or shelter, provision of poor quality nutrient, etc. Furthermore, it also had some social impacts such as many families were forced to live apart as men and young boys stayed in camp with their livestock and women were left behind in villages to take care of their families. Similar social impacts were reported in a case study by World Society for the Protection of Animals.

Compensations to agricultural losses: Government provided assistance or relief fund to support horticultural- and agricultural-crop loss due to drought. In the year 2012–2013, government provided USD 147.2 and 55.2 per hectare as a compensation for the horticultural and agricultural crops damaged by drought respectively. Also, farmers from DPA bought crop insurance. In case of worst and prolonged drought, the government also provides loan, subsidy scheme, waives interest and increases loan repayment period as relief measures to the farmers. According to Government of Maharashtra, total drought damage compensation cost estimated for the year 2012–2013 was approximately USD 133 million. This relief measure had some negative impacts such as some farmers did not pay back loans in spite of their ability, and waited for crop loan and interest waiver from the government. A similar case was observed in case of electricity bill payment. On an average, the level of satisfaction from this relief and mitigation efforts was very less among the respondents on 5-point Likert type response.

Provision of employment scheme, cattle camps, fodder depots, and water supply tankers; and compensations for damaged agricultural crops and fruit gardens were the major drought mitigation measures found in the Bhima catchment. According to Drought Memorandum, Government of Maharashtra, these mitigation measures (except NREGA) incurred cost of approximately USD 332 million for drought 2012–2013. Even though these administrative drought mitigation measures have provided some relief to the affected households, the level of satisfaction was still low among them. Only few farmers got information and benefit from contingency plan. There was lack of awareness and information on administrative drought mitigation measures observed amongst potential beneficiaries. Also, according to the Indian Agriculture Ministry's crisis management plan, drought is not a disaster, it is a management issue; improvement in the management of these relief measures is necessary to mitigate the drought impacts on farming community.

DROUGHT MANAGEMENT FRAMEWORK IN INDIA

The Government of India has devised many short-, medium- and long-term strategies to mitigate and overcome adverse effects of drought, and has implemented relief and development programmes in cooperation with the concerned states. These measures include ensuring availability of food grains and fodder, judicious use of surface and groundwater, prevention of migration of cattle camps, appropriate selection of crops, cropping sequences and agronomic practices, promotion subsidiary income of the affected people, and employment generation in rural areas.

Among these, selection of crops, cropping sequence and agronomic practices appropriate for drought play a very crucial role to reduce its effect on food grain production of the country. To assist farmers of the country for this endeavour, Indian Council of Agricultural Research has accomplished a huge task of designing drought plans for every district of the country. These plans have been shared with the respective state governments and district administrations for popularization among farmers. Nationwide demonstrations of these contingency plans have been taken up under its mega project on National Initiative on Climate Resilient Technologies (NICRA). The district wise plans are available at http://www.crida.in:82/contingencyplanning/ and all the line departments must make use of these plans to reduce drought effect on farming community.

CONCLUSION

Drought will be a part of Indian agriculture. It has considerable socio- economic effect at macro and micro level. To reduce its impact on livelihood of farmers adaptation and mitigation measures at farmers as well as government levlel are required. This include diversification of system, storage of food and feed, at farmers' level and mitigatigation measures like employment, drinking water supply, cattle camps, compensation schemes etc. are initiated. However, long term measures are required to combat drought.

REFERENCES

Bandopadhya J 1988 Water scarcity: by choice. *Seminar*, **346** 24-27

Benson C and Clay JE 1998 The impact of drought on Sub-Saharan African economies: a preliminary examination, World Bank Technical Paper 401. Washington, D.C. World Bank.

Bidinger PD, Walker TS, Sarkar B, Ram Murthy and Babu P 1991 Consequences of mid 1980 drought: longitudinal evidence from Mahboobnagar, *Economic and political weekly* **26** 105-A114

Biradar Nagaratna and Sridhar K 2009 Consequences of 2003 Drought in Karnataka with Particular Reference to Livestock and Fodder, *Indian Journal of Human Ecology* **26(2)** 123-130

DARE, 2003. Drought 2002 states report-AP, Karnataka and Tamil Nadu, Department of Agriculture Research and Education, Ministry of Agriculture, Government of India, New Delhi

NSSO 1999 *Common Property Resources in India: NSS 54th Round, January 1998 – June 1998*. New Delhi: National Sample Survey Organisation, Department of Statistics and Programme Implementation, Government of India

Palmer WC 1965 Meteorological drought. Research paper no.45, Office of climatology, U.S. Weather Bureau, Washington D.C.

Pandey S and Bhandari H 2006 Drought perpetuates poverty, *Rice Today*, April-June, 37 pages

Rathore MS 2005 State level analysis of drought policies and impacts in Rajasthan, India. Colombo, Sri Lanka: IWMI. 40p. (Working paper 93 : Drought Series Paper No. 6)

Ravindranath M, Venkaiah K, Rao MN, Arlappa N and Reddy CG 2005 Effect of drought on nutritional status of rural community in Karnataka, *Journal of Human Ecology* **18(3)** 245-252

Roy AK and Hirway I 2007 Multiple impacts of droughts and assessment of drought policy in major drought prone states in India. Gujarat, India: Centre for Development Alternatives

Sen A 1981 Poverty and Famines: An Essay on Entitlement and Deprivation. Clarendon Press, Oxford

Udmale P, Ichikawa Y, Manandhar S, Ishidaira H and Kiem AS 2014 Farmers' perception of drought impacts, local adaptation and administrative mitigation measures in Maharashtra state, India, *Indian Journal of disaster risk reduction* **10** 250-269

Wood AP 1976 Farmers response to drought in Ethiopia. In Hussein, A. M. (ed.), Rehab: Drought and Famine in Ethiopia. IAI, London.

24

Ensuring Livelihood Security during Drought

Suresh Kumar

INTRODUCTION

Ensuring livelihood for survival of human being and livestock in perpetually low rainfall areas is a challenge, more so during drought, in view of low production potential, inaccessibility, poor infrastructure and inequity. Attempts to provide secured livelihood has attracted the attention of planners since long and it continues to be so by embracing the advancements in research and developments in science and technology. These approaches, their usefulness in tackling the issue of livelihood and the future strategies are discussed in this chapter.

HISTORICAL PERSPECTIVE

Ensuring livelihood through drought relief has been continuing in India for many years, though systematic efforts to deal with the drought affected areas did not exist in pre British India. The British administration adopted famine codes on the basis of recommendations of the Indian Famine Commission in 1880. A systematic approach, based on climatic, economic and social factors, was therefore employed for the first time to respond to famines in this famine code. But it failed to stop famine deaths nor did they eliminate large scale famines in the pre-independence era. This was due to the fact that such interventions were not automatic and in fact more of exploitative of job seekers in public relief works. Most importantly, government did not stop hoarding of food grains, resulting in their price rise. Consequently, poor and marginalized could not afford to purchase mere necessity of food. Thus, the famine codes meant to minimize famine-related deaths failed to do so. This has been ascribed to late implementation of famine relief after drought was well advanced.

Principal Scientist & Ex Head
Central Arid Zone Research Institute, Jodhpur 342003 (Rajasthan) India
Email: sk_ecology@yahoo.co.in

SCARCITY VERSUS FAMINE

After independence, attempts were made to differentiate scarcity from famine. Famine means acute distress, while scarcity happens to be the beginning of distress. The famine relief approach became scarcity relief firstly during drought of 1965-66, when aim was to prevent starvation deaths rather than to minimize the death toll. Physical access as also economic access to food became primary focus to help affected population to ensure prevention of drought-related starvation deaths by way of free kitchens, free relief, and opening relief works. Even this was essentially an ad-hoc approach i.e., providing relief and ensuring livelihood when drought had actually occurred and advanced.

DROUGHT RELIEF FOR DEVELOPMENT

Hence, it was thought appropriate to integrate relief efforts with development programs, so as to maximize development from relief expenditures. It was aimed to have preparedness measures such as the establishment of a national food security system and perfect the monitoring techniques for crop conditions and reservoir levels. These development programs reduced vulnerability to drought by enhancing the coping capacity in drought prone areas. As the development status of a society enhances, they become less and less dependent on governmental relief programs.

ENSURING LIVELIHOOD AND INCREASING RESILIENCE

Basically, ensuring livelihood has been linked with and part of drought management practices in India. It has six components: (1) Operationalization of an early warning system; (2) drought preparedness measures; (3) conserving water; (4) managing resources for stabilizing crop production; (5) assuring physical and economic access to food; and (6) preserving farmers' assets and enabling to recover quickly when the normal monsoon returns. Efforts are directed to increase resilience in coping drought with minimum damage to natural resource base. Many of these objectives can be met in Government sponsored schemes such as the National Rural Employment Programme and the various rural infrastructures development programs. These give employment to the rural work force to stabilize incomes. Large number of developmental activities/ works can be taken up as per local needs. Such works should be listed in advance in preparation of drought management. Prior evaluation of these works in the basket of relief in advance helps to select the appropriate works that provide jobs to the drought- affected population on short notice. Thus, besides providing cash to increase purchasing power of people, assets are also built in the affected areas to improve their resistance to drought. Though physical access to food is ensured in the National Food Distribution System, employment generation schemes provide economic access to food. Contingency crop plans are prepared to meet exigencies created by the erratic monsoon behavior. These include keeping buffers stocks of seeds to enable farmers to switch to other crops. Arrangements for resource input and energy support such as diesel and power supply, are important and are essentially required to stabilize farm - level crop production, as for as possible.

WATER CONSERVATION

In order to minimize crop loss, water available in reservoirs and from exploitation of surface and groundwater needs to be assessed to meet the critical needs of crops under moisture stress conditions. Water budgeting to ensure optimum use of scarce water resources by prioritizing the uses is the only way to achieve it. Various stakeholders interests like agriculture, public health, engineering, groundwater reservoir operations, and the State Electricity Board should establish that drinking water is the first water use priority, followed by fodder programs and finally irrigation and industry. National water policy (2012) and state water policy (2010) also addressed this issue.

STABILIZING CROP PRODUCTION

Extensive research (reviewed in Garg and Burman 2011) has established that drought effects in arid legumes could be alleviated through phosphorus application that mediates the chain of physiological and biological processes to impart stability. Similar effects of applied potassium are also seen. Adequate nutrient management can thus help to significantly improve the yield even under drought stress. Yet, crop failures during drought is a common scenario. Crop failure in drought reduces the income of farmers drastically. Even the labour that depends on agricultural operation is under distress. Thus employment generation program assumes significance and priority in the drought management strategy. This assures purchasing power in the hands of the affected population, as it provides economic access to food and enables rural households to preserve productive assets like cattle and land. This also prevents or minimizes migration. Construction Programmes related to water conservation, small- scale irrigation, soil conservation, and afforestation programs have been successful in the past in the drought-affected areas.

STABILIZING GRAZING/ RANGE LANDS PRODUCTION

Since large majority of population is pastoralists in desert, decline in availability of grazable material due to drought adversely affect their survival. The extent and depth of severity becomes evident if we compare income from different subsectors to total income of a household in western Rajasthan. Nearly 59-71% of income is earned through agriculture (rainfed + irrigated + horticulture) while 28 to 42% from livestock rearing and 14 to 24% from household industry (Kar 2014). With drought affecting agricultural income of 59 to 71% to as low as nil, the livestock and household industry become mainstay of the desert dwellers. In fact mean per unit income from bovine is far better than from irrigated croplands (Table 1). Such trends are seen in other deserts of the world such as South Africa (Vetter 2009). In order to cope with drought impacts, these rural ecosystems need to be robust and resilient. Berkes (2007) defined resilience as a capacity to absorb change and adapt. This involves resilience in production system, adapting to earn income from different locations by diversifying income streams, institutional support and access to resources. Production system in Indian arid zone though rain fed are truly resilient: a mixture of 2-3 crops, with perennial as well as annual palatable grasses and herbs, browse shrubs and trees. Even if one crop fails another crop does give some yield, besides getting fodder from grasses, forbs, shrubs and trees. These systems have evolved over centuries though components in different rainfall zones vary.

Provisioning diversification in income earning opportunities is yet another area that requires attention. "Skill India" scheme of Government of India might as well work wonders here in this area. Institutional support and access to resources are the two issues being addressed in various welfare schemes of central and state government. Those families having above features and diversified income sources will have much less vulnerability to impacts of drought, especially in terms of five classes of livelihood assets, which are used to measure vulnerability in the framework of sustainable livelihood (Chamber and Convey 1992). These are:

1. Human capital such as health, education and skills
2. Natural capital i.e. access to ecosystem services and products
3. Social capital, including social and kinship networks and support
4. Physical capital, such as infrastructure and housing
5. Economic capital, i.e. income, savings and access to credit.

Thus capacity building of land users becomes important to empower them with these five capitals to be able to cope with drought onslaught. Though strength in last four capitals will ultimately build up human capital, it is the natural capital that plays a major role in drought coping and providing livelihood. A large body of knowledge from decades of research in Indian arid zone has conclusively shown that building natural capital is possible in degraded to severely degraded landscape, the time to recovery being proportional to degradedness of the land patch.

Even in the extremely degraded, barren landscapes, there have been many success stories of re-gaining the productive phase through ecological rehabilitation, many of which are described in this book in different chapters. The missing link however is regulation of resource use, this being a common property especially in grazing land areas. As such, there is neither a grazing policy at central or state level, nor is there a community enforced regulation of access and use of these resources. Consequently, in the events of drought, people become so acutely dependent on state help for supply of forage for livestock that huge amount of money is spent on its transportation alone, leave aside its cost. There is an example of 2002-03 drought in Shergarh Block headquarter in Jodhpur district where community land fenced and reseeded by Forest Department was able to meet all the forage need of local livestock without any import from outside . Such "island success stories" are pointer to the way in which policy thought process need to go for. The policy cue is to focus development of vast rangeland which occur mostly on wastelands occupying an area of over 36% including 20% culturable wastes so that grazable material for livestock become available to be stocked like food grains to meet forage demand during drought. A policy is also needed to develop "livestock feed banks" or "forage banks" for meeting the immediate needs of fodder. Support is also needed to link livestock production system with commercialized dairy and livestock based industries with related infrastructure in place to ultimately have "special dairy zone" (SDZ) on the tunes of "SEZ".

Table 1. Average per unit income from different sub- sectors of agriculture to the total agricultural income in the agro-climatic zones (ACZ) (Kar 2014)

ACZ	Rainfed land (Rs./ha)	Irrigated land (Rs./ha)	Horticulture (Rs./ha)	Bovine (Rs./ animal)	Ovine (Rs./ animal)	Household industry (Rs./worker)
ACZ-I	7.42	22.62	21.57	33.26	0.56	14.57
ACZ-II	2.71	34.71	15.62	22.22	0.74	24.01
ACZ-III	7.30	22.34	19.68	35.47	0.83	14.38
ACZ-IV	6.42	25.63	22.68	28.89	0.68	15.70
Mean	5.858	27.19	19.99	29.80	0.73	16.44

DEVELOPMENTAL ALTERNATIVES

Digging of tube wells, ponds, field channels, soil conservation and water harvesting works and road construction in areas where road links do not exist are prioritized activities in drought relief works that also provide livelihood besides creating assets which go a long way in drought proofing in future. Preservation of assets in drought affected areas by assured income through employment generation programmes helps to resume normal economic activities at the on-set of the normal monsoon season. Normally, drought results in massive depletion of farm assets through sale and death of cattle, which takes a lot of time to recover. Saving livestock in drought areas by increasing the availability of fodder by bringing it from surplus areas at appropriate time requires critical assessment and decision making. In fact raising of fodder crops by providing fiscal and technical inputs in drought affected areas receives priority by Agricultural Department which often distributes seed kits of forage species. In all these programmes community participation always plays a crucial and constructive role.

COMMUNITY PARTICIPATION

If resources are diverted to totally unproductive uses, it will be detrimental in the long run, because the resources spent on unproductive forms of relief will drain resources needed for the country's developmental needs and, consequently, reduce its ability to manage future droughts. Keeping this in view, creating productive assets like small -scale irrigation and moisture conservation works in relief operations is preferred as in the long run, this enables the people better withstand droughts in the future. Maru Gochar Yojana proposed for development of community grazing lands in western Rajasthan involving NGOs and communities is another ambitious initiative of government which conserves natural resources, makes available the fodder at the critical scarcity time, ensures livestock survival and prevent migration. Such employment needs to be provided within manageable distance from the residents of the affected area. As projects are scattered involving various departments, these works are not completed resulting in a decline in the productive capability of these projects. Consequently, much to the dismay of authorities these employment generation works became an instrument of relief provision rather than a measure that creates durable assets, as originally intended. Further, the relief provision involves resource transfers in the form of income, food, water, and fodder resources from surplus to deficit areas. This resource transfer approach increasingly make people dependent on state rather than supporting their inherent potential to cope with short-term climatic variation.

DROUGHT MANAGEMENT STRATEGY

Alternative drought management strategy should therefore gradually eliminate or reduce relief and increase the capacity of the affected areas to withstand droughts on their own. Droughts are common and its recurrence is a certainty. But we respond to drought considering it as a transient phenomenon, and consequently all response efforts are directed to meet single-season disturbances. The employment generation program helps able-bodied persons; supplementary nutrition programs for pregnant women, lactating mothers, and children; gratuitous relief for infirm and physically handicapped people; cattle camps for cattle; and so forth. In fact people in drought-prone regions have evolved a diversified occupational pattern to deal with the crisis created by drought situations. These include short- term and long-term adjustments to handle both good and bad years. These adjustments are not mainstreamed by these interventions, because these isolate indigenous people's inherent coping capacity to drought in their socio-economic fabric. Monsoon management approach to keep their livelihood system adapted to climatic changes is their biggest strategy by keeping reserves and diversifying economically. Subbiah (1993) has outlined these as "...traditional strategies rely on primary, secondary, and tertiary production systems of dry zones to cope with droughts and seasonal fluctuations; these strategies have viable ingredients compatible with the environment.

Response systems have mechanisms to absorb the shock of drought and reduce drought impacts. Although the primary production system (such as crops) may fail secondary production systems (such as livestock) reduce sensitivity. The tertiary production system includes off-farm employment opportunities that provide near- immunity to the impacts of drought...". Though some of these may have lost pertinence in existing era of technical advances and increasing market forces, yet it is essential to understand these systems so as to revitalize and strengthen them for the benefit of the farmers.

Since there is a limit to primary and secondary production systems, it can only provide limited employment in the drought in dry zones without affecting natural resource base. Thus it is essential to create more of non-farm jobs in the agro-based village industries and even tourism, so as to provide financial security against droughts. At the tertiary level, promoting local skills such as handloom work, carpet weaving, handicrafts, and small-scale fabricators goes a long way in drought coping.

Thus monsoon management should successfully increase primary and secondary production systems. The tertiary production systems, should transform current drought management methods from resource transfer to resource regeneration. Some of the research mediated off-farm vocations deserving promotion in both normal and drought years are discussed below.

1. Solar candle making machine (1060x750x200mm) uses 18kg wax to melt under solar heat and put in caveats to make candle. It has an outer box, melting chamber, glazing, kamani, tilting controller and outlet. It helps conserve conventional fuel, is safe and convenient. It costs Rs. 15000/- per unit and payback period is 2-3 years.
2. Solar cooker (960x320x210mm) can be manufactured as per CAZRI drawings and specification. These are cheaper, safe, and generate employment for carpenter on one hand while save fuel wood on the other hand.
3. Novel hand hoe with a hole in traditional kassi is being favoured for weeding by villagers as it requires much less power, reduces drudgery and increases crop yield. Blacksmiths in the village can adopt it for local manufacturing year round.
4. Manufacturing solar cooker for animal feed generates employment and saves fuel while giving nutrition to livestock.
5. Dried vegetables and fruits are commonly kept in each household to meet the demand during scarcity. It is in fact, an industry in western Rajasthan. This can be further augmented by solar dryers to save conventional fuel and minimize post harvest losses.
6. Production of gum arabic from *Acacia senegal* through ethephon induction is another additional income generating activity in the dry

areas. With Rs 10/- per tree invested, up to 500g of gum is produced from each tree and that earns an income of Rs 150/- to 300/- per tree depending on quality. Planting more of gum producing trees of the like will ensure future availability of gum during drought.

7. Making of supplement feed block at community level will help get employment for people and feed blocks for the livestock. While feed block improves productivity it also increases milk yield by balancing the nutrient uptake (Narain and Kar 2005).

8. Land shaping for *in situ* water conservation, will also generate substantial employment. These include circular catchments around perennial species like ber, half-moon terraces on sloppy surface, vegetative barriers of economically important under shrubs like *Aloe vera, Senna angustifolia, Ziziphus, nummularia, Calligonum polygonoides* and *Panicum turgidum, Lasiurus sindicus* and *Cenchrus ciliaris*. Besides all soil and water conservation works fall in this category.

9. Underutilized Plants for Life Support System: In a survey conducted in four districts namely Jodhpur, Bikaner, Barmer and Jaisalmer, 51 plants were found to have many uses for their survival and sustenance. Stems of 28 species are used in furniture, cartwheels, tools, stilts, windows doors, toys and utensils, fencing, thatching, ropes, fuel, fodder, brooms, matchsticks, musical instruments, incense, deodorants, toothbrushes, dying leather, worshiping, magicoreligious and explosives. Leaves of 17 taxa are used as fodder, vegetable, preparing drinks, mosquito repellant, pesticides, perfumes etc. Some 13 taxa are used as whole plant for fuel, fodder, food, fencing, thatching and even salt making (*Lycium barbatum*). Fruits of 8 taxa are used as vegetable, pickles or eaten raw. Seeds of 10 taxa are used as pickles, delicacies vegetables, scarcity foods, pesticides and magicoreligious beliefs. Roots of 6 taxa are used in making spoon, dying of leather. Gum of 4 taxa has neutraceauticals value while other 4 taxa have uses in stuffing pillows and other uses. These species have multiple uses. *Prosopis cineraria* and *Calotropis procera* each had 17 uses followed by 16 uses of *Capparis decidua*, 13 of *Ziziphus nummularia*, 9 of *Calligonum polygonoides*, 8 of *Prosopis juliflora*, 7 of *Tecomella undulata, Acacia nilotica, A. senegal* and *Leptadenia pyrotechnica* (Kumar 2004).

10. Medicinal plants sourced from natural habitats: Some 46 species found in high trade are collected from wastelands (Ved and Goraya 2007). Although these don't require immediate attention as these are abundantly available, their domestication and cultivation is essential to have uniform quality of the produce. These include (Kumar et al 2005) the following:

Abrus precatorius (Rati), *Achyranthes aspera* (Apamarg), *Aerva lanata* (Cheroola), *Andrographis paniculata* (Kalmegh), *Boerhavia diffusa* (Punarnava),

Citrullus colocynthes (Indrayan), *Convolvulus microphyllus* (Shankhapushpi), *Cyndon dactylon* (Ourva), *Cyperus rotundus* (Nagar motha), *Ocimum americanum* (Ban tulsi), *Peganum harmala* (Harmal), *Pluchea lanceolata* (Rasna), *Sida cordifolia* (Bala), *Sisymbrium irio* (Khubkalan), *Tephrosia purpurea* (Sarpankha), *Vetiveria zizanoides* (Lavancha), and *Withania coagulens* (Panirdodi).

Evidently, enough scope exists in drylands and deserts to cope with drought by strengthening existing livelihood systems and providing desired marketing support to sustain these income generating activities.

RECENT DEVELOPMENT IN DROUGHT MANAGEMENT: NATIONAL SCENARIO

National Disaster Management Authority, Govt. of India (NDMA) has prepared a National Disaster Management Plan. Drought is recognized as a major disaster by NDMA. In addition to guidelines for drought management, NDMA has highlighted the need for preparing vulnerability maps, refining assessment, monitoring and early warning system, improving criteria and key indicators of drought declaration, providing guidelines for hazard risk vulnerability assessment and promoting research related to water conservation and agriculture. For achieving success, interagency coordination has been stressed on overall disaster governance, response, warnings, information exchange and in revision of rules, norms, codes, standard and laws for regulation and policies. Besides, drought mitigation by way of developing storage facilities for water, feed and food has been aimed to be achieved by way of water conservation structures, rainwater harvesting in social housing schemes, watershed development projects, scientific dryland farming and public private partnership. Ensuring availability of agricultural credit, agri-inputs, finance, marketing facilities and crop insurance will further mitigate the risks. For success in drought mitigation, the role of training and capacity building, inclusion of drought management in university curricula, awareness generation and empowering women, marginalised communities and differently abled persons has also been duly recognized (www.NDMA.N.D.M.P).

REFERENCES

Berkes F 2007 Understanding uncertainty and reducing vulnerability: lessons from resilience thinking, *Nat. Hazards* **41** 283-295

Chambers R and Conway G 1992 Sustainable rural livelihoods: practical concepts for 21st Century, IDS discussion paper **296**. IDS. Brighton.

Garg BK and Burman U 2011 Physiological basis of yield improvement of arid zone crops through nutrient management, *Plant Stress* **5** 73-81.

Kar A 2014 Agriculture landuse in arid western Rajasthan: Resource exploitation and emerging issues, *Agropedology* **24(2)** 179-196

Kumar S 2004 Traditional Ethanomedicinal Plants of the Indian Thar: Their Status in Nature and Possibilities of Cultivation and Trading, Report submitted to DST, New Delhi

Kumar S, Parveen F, Goyal S and Chauhan A 2005 Trading of Ethnomedicinal plants in the Indian Arid Zone, *Indian Forester* **131(3)** (March Issue): 371-378.

Narain Pratap and Kar Amal 2005 (Eds.) Drought in western Rajasthan: Impact, Coping Mechanism and Management Strategies. Arid Agro-ecosystem Director, National Agricultural Technology Project, CAZRI, Jodhpur, India.

NDMA (National Disaster Management Plan) 2016 A publication of National Disaster Management Authority, Govt. of India, May 2016 New Delhi (PBF),

National Water Policy, 2012 Govt. of India, Ministry of Water Resources. wrmin.nic.in

State Water Policy, 2010 archieve.india.gov.in

Subbiah AR 1993 Indian Drought Management: From Vulnerability to Resilience. In: Wilhite, D. A. (Ed.) Drought Assessment, Management and Planning: Theory and Case Studies. Kluwer Academic Publishers, Boston, MA.

Ved DK and Goraya GS 2007 *Demands and Supply of Medicinal Plants in India.* Executive summary. National Medicinal Plant Board, New Delhi & FRLHT, Bangalore. pp.8

Vetter S 2009 Drought, change and resilience in South Africa's arid and semi-arid rangelands, *South African Journal of Sciences* **105** 29-3.